BIOMEDICAL RESEARCH

STRATEGIC PLANNING FOR SUCCESS

BIOMEDICAL RESEARCH

STRATEGIC PLANNING FOR SUCCESS

Y. ROBERT LI
WITH
MICHAEL P. MAHALIK
AND
JOHN M. KAUFFMAN, JR.

New York

For permission to use material from this book please contact us:
Telephone 631-231-7269; Fax 631-231-8175
Web Site: http://www.novapublishers.com

Library of Congress Cataloging-in-Publication Data

Library of Congress Control Number: 2013940378

ISBN: 978-1-62808-105-3

Published by Nova Science Publishers, Inc. † New York

This book is dedicated to the newly established Campbell University School of Osteopathic Medicine

Contents

Preface

Research is generally regarded as a core element of the mission of institutions of higher education. The extent to which higher education institutions are engaged in research and development activities has a key role in determining the status and quality of these institutions and the contributions which they make to economic and social development. While there are books addressing the overall strategies on how to develop world-class research universities, a reference book dedicated to the discussion of the practical approaches to developing biomedical research in traditionally non-research-intensive institutions of higher education is currently not available. Such a reference book is of urgent need because (i) biomedical research has been receiving unprecedented attention and support from both academics and administrative professionals at thousands of traditionally non-research-intensive institutions of higher education throughout the world; and (ii) academics and administrative professionals at traditionally non-research-intensive institutions are facing ever-increasing challenges regarding development of competitive biomedical research programs using limited amounts of resources to increase their institution's prestige and value. This book is intended to serve as a general reference for developing strategic biomedical research in medical schools (including both allopathic and osteopathic medical schools) and other colleges of health professions (such as schools of pharmacy, nursing, and physical therapy) located at or affiliated with traditionally non-research-intensive universities. The book focuses on discussing the importance of biomedical research in advancing the prestige and increasing the value of such institutions and details the key strategies for establishing and maintaining nationally and internationally recognized biomedical research programs. This book is divided into four units, as listed below.

Unit I: Introduction to Biomedical Research
Unit II: Structural Components of Strategic Biomedical Research
Unit III: Supportive Offices and Committees for Strategic Biomedical Research
Unit IV: Strategic Biomedical Research Planning

To set the stage for the subsequent discussion of the structural components of strategic biomedical research and the practical approaches to developing institutional biomedical research, Unit I provides two introductory chapters. Chapter 1 defines biomedical research and its scope, research process, and research ethics. Chapter 2 discusses the essential nature of biomedical research, including its impact on public health and economic development in a broad manner and on the prestige and value of the individual institutions of higher education.

Unit II consists of three chapters (Chapters 3-5) devoted to the discussion of the physical structural components of strategic biomedical research. Chapter 3 describes common institutional core biomedical research facilities and introduces the core facilities at several prestigious medical institutions. This chapter is intended to serve as a general reference for constructing core biomedical research facilities at traditionally non-research-intensive institutions. Chapter 4 covers the definitions and basic concepts related to laboratory animal research and focuses on discussing the principles of design and construction of laboratory animal facilities as well as facility accreditation. The policies and guidelines on humane care and use of laboratory animals are covered in detail in Chapter 7 of Unit III. The last chapter of Unit II provides an overview of the definition, classification, and general features of biomedical research laboratories, and introduces the principles of planning and designing of the research laboratories.

Unit III contains three chapters (Chapters 6-8) that cover the institutional supportive offices and committees for strategic biomedical research. Chapter 6 describes the duties of the office of contracts and grants administration and its relationship to the department of research administration. The chapter also introduces the most current (2012) National Institutes of Health (NIH) grant policy statement as well as common terminology on grant application and administration. The chapter ends with an overview of the NIH grants process. Chapter 7, as noted above, covers the current policies and guidelines on humane care and use of animals in research with a focus on the responsibilities and constitution of the Institutional Animal Care and Use Committee (IACUC). This chapter also describes animal welfare assurance as well as the United States government principles for the utilization and care of vertebrate animals used in testing, research, and training. Chapter 8 of Unit III discusses the basic ethical principles and federal regulations of human subjects research with an emphasis on the role of Institutional Review Board (IRB) in the protection of human subjects in biomedical research.

Unit IV consists of three chapters (Chapters 9-11) on strategic biomedical research planning. Chapter 9 introduces the three-set-factor model for establishing world-class research universities. The chapter also discusses the role of this model in guiding the development of strategic biomedical research at both research universities and traditionally non-research-intensive institutions. Chapter 10 defines institutional vision and mission statements and discusses ways to develop effective vision and mission statements on strategic biomedical research. This chapter addresses three key areas (who we are, what we do, and why we do it) of developing a compelling mission statement on strategic biomedical research at institutions of higher education, especially osteopathic medical schools. Chapter 11, the last chapter of the book, describes five major biomedical research goals and the corresponding strategies to meet the goals and focuses on discussing their relevance to developing strategic biomedical research at traditionally non-research-intensive institutions. The five major strategic goals are (i) allocating resources and building research infrastructure to optimize resource utilization, (ii) recruiting and retaining research-active and multitasking faculty to secure academic talent, (iii) defining biomedical research areas and priorities to build institutional signature research, (iv) stimulating biomedical research and scholarly productivity of faculty to maximize research output and investment return, and (v) fostering interdisciplinary programs to unify university wide research enterprise.

This book is based on the authors' professional experience in developing biomedical research programs at various institutions of higher education, discussion with peers (both national and international) at respected research universities as well as non-research-intensive institutions with ongoing biomedical research programs, and a critical review of the current literature. It should be noted that the statements in this book are based on the integrated information from multiple respected sources of high quality combined with the authors' own understanding of the field and direct professional experience. The major references cited are listed at the end of each chapter. It is hoped that this book will be used as a guide for those who are involved in developing institutional strategic biomedical research to benefit society as well as to increase the institution's prestige and value. It is also hoped that the book will provide a useful reference for the general public to gain a quick and thorough understanding of biomedical research and how to develop biomedical research, a main domain of the 21st century scientific research enterprises.

Y. Robert Li, MD, MPH, PhD
Professor and Chair of Department of Pharmacology
Assistant Dean for Biomedical Research
Campbell University School of Osteopathic Medicine
Adjunct Professor of Virginia Polytechnic Institute and State University

with

Michael P. Mahalik, PhD
Senior Associate Dean for Academic Affairs and Research
Professor of Pharmacology
Campbell University School of Osteopathic Medicine

and

John M. Kauffman, Jr., DO, FACOI, FACP
Dean and Chief Academic Officer
Professor of Internal Medicine
Campbell University School of Osteopathic Medicine

Buies Creek, North Carolina
March 2013

Unit I.
Introduction to Biomedical Research

Chapter 1

Defining Biomedical Research

Abstract

Biomedical research is one of the main domains of scientific research and often serves as a core for multidisciplinary collaborative research endeavors at institutions of higher education. Biomedical research is essential to the advancement of medicine and improvement of human health. This chapter defines biomedical research and its scope and process. It also introduces the fundamental principles of biomedical research ethics and the commonly shared values of the responsible conduct of research. It is hoped that this chapter will provide the readers a big picture on biomedical research and lay a foundation for understanding the structural components of strategic biomedical research and research development.

Chapter at a Glance

1. Overview

Research is generally regarded as a core element of the mission of institutions of higher education. The extent to which universities are engaged in research and development activities has a key role in determining the status and quality of these institutions and the contribution which they make to economic and social development [1]. This chapter first defines research and introduces the general scheme for classification of research. It then defines the commonly encountered terms, including biomedicine, biomedical science, and biomedical research, and discusses the challenges and controversies regarding the terminology and scope of biomedical research. The chapter also examines biomedical research process and introduces the basic principles of research ethics and responsible conduct of biomedical research. This will lay a foundation for the subsequent coverage of the structural components of strategic biomedical research as well as the practical approaches to developing biomedical research at institutions of higher education, especially traditionally non-research-intensive universities.

2. Defining Research

In the broadest sense of the word, the definition of research includes any gathering of data, information, and facts for the advancement of knowledge [2]. Research takes different forms, including:

- Scientific research
- Humanity research (study of the human condition)
- Artistic research (combining some of the procedures of scientific research and artistic creation)

The strict definition of scientific research is using scientific methods to perform a detailed study of a subject, especially in order to discover new information or reach a new understanding. Finding a definitive answer is the central goal of scientific research [2]. Scientific research provides information and theories for the explanation of nature including humans and the properties of the world.

Scientific research must be systematic and follow a series of steps and a rigid standard protocol. These rules are broadly similar but may vary slightly between the different fields of science. The major steps in conducting scientific research include the following:

- Identifying a scientific research problem
- Searching and reviewing the literature
- Specifying the purpose of the scientific research
- Determining specific scientific research questions or hypotheses
- Collecting data using scientific methods
- Analyzing and interpreting the scientific data
- Reporting and evaluating the scientific research

Scientific research can be divided into different categories according to their academic and application disciplines. A common scheme classifies scientific research into two broad categories: (i) basic scientific research and (ii) applied scientific research, with the distinction between them being stated many decades ago by the director of the United States Office of Scientific Development and Research (**Box 1-1**).

Box 1-1. Definitions of basic and applied scientific research

In 1945, when the director of the United States Office of Scientific Development and Research, Vannevar Bush, proposed the establishment of the National Science Foundation (NSF), he made the following distinction between basic research and applied research [3]:

- Basic research is performed without thought of practical ends.
- Basic research results in general knowledge and an understanding of nature and its laws.
- This general knowledge provides the means of answering a large number of important practical problems, though it may not give a complete specific answer to any one of them.
- The function of applied research is to provide such complete answers.

The NSF definition thus identifies the main objective of basic research as the acquisition of knowledge without the obligation to apply it to practical ends.

In his 2012 Science article entitled, "NIH Basics", the United States National Institutes of Health (NIH) director, Francis S. Collins, defines basic research as systematic study directed toward fuller knowledge or understanding of the fundamental aspects of phenomena and of observable facts without specific applications in mind. Since 2003, the proportion of NIH funds spent on basic research, defined in this way, has ranged from 53 to 57%, standing at 54% for fiscal year 2012. Basic research also accounts for most of the 135 Nobel Prizes won by NIH-supported scientists, including the 2011 awards to Bruce A. Beutler and Jules A. Hoffmann for their discoveries about innate immunity, and the late Ralph M. Steinman for adaptive immunity [4].

2.1. Basic Scientific Research

Basic research, also known as fundamental or pure research, is driven by a scientist's curiosity or interest in a scientific question. This type of research is often purely theoretical with the intent of exploring the unknown and expanding man's knowledge, rather than to create or invent something of immediate commercial value. Hence, in general, there is no obvious commercial value to the discoveries that result from basic scientific research. Examples of basic scientific research include:

- Investigation of a specific function of a cellular protein
- Investigation of a specific intracellular signaling pathway leading to uncontrolled cell proliferation

- Investigation of the origin of life
- Investigation of the composition of neutrons

It is widely appreciated that a basic, fundamental understanding of all branches of science is needed in order for progress to take place. In other words, basic scientific research lays down the foundation for the applied science that follows. If basic work is done first, then applied spin-offs often eventually result from this research. As the Nobel Laureate George F. Smoot (2006 Physics) of the University of California at Berkeley and the Lawrence Berkeley National Laboratory says, "People cannot foresee the future well enough to predict what's going to develop from basic research. If we only did applied research, we would still be making better spears [5]."

The fundamental role of basic research in advancing medicine has been demonstrated through the history of modern medicine. In the Congressional Justification of the NIH fiscal year 2013 budget request [6], the NIH director Francis S. Collins states that basic biomedical research makes it possible to understand the cause of disease onset and progression - an understanding that is essential to the development of better diagnostics, the design of preventive interventions, and the discovery of new treatments and cures. For instance, basic biomedical research on the communities of bacteria and other microorganisms (the "microbiome") that normally live in the human digestive tract has revealed that these organisms play an essential role in maintaining health by aiding digestion. Researchers have now uncovered evidence that changes in the composition and activity of the microbiome contribute to the development of obesity, which has reached epidemic proportions in the United States as well as many other developed and developing countries. It is hypothesized that the normal bacterial communities are being affected by environmental factors such as diet and antibiotic exposure, and that these changes can be passed on from mother to child. Such basic research offers great promise to provide novel approaches to the prevention of obesity [6].

2.2. Applied Scientific Research

Applied scientific research is designed to solve practical problems of the modern world, rather than to obtain knowledge for knowledge's sake [7]. As noted above, basic research, however, provides a foundation for applied scientific research. Examples of applied scientific research include:

- Investigation of ways to augment agricultural crop production
- Investigation of ways to treat or prevent a specific disease
- Investigation of ways to improve the energy efficiency of cars

It has been argued that the time has come for a shift in emphasis away from purely basic scientific research and toward applied science. This trend is necessitated by the ever-increasing public health, social, and economic problems resulting from global overpopulation, pollution, and the overuse of the earth's natural resources. However, such a shift does not and should not undermine the efforts putting into basic scientific research, especially in the field of biomedical science. In this context, as discussed next, basic biomedical research provides an indispensible foundation for translational and clinical research to build on to eventually lead to the improvement of the health of the human population.

3. Defining Biomedical Research

3.1. Defining Biomedicine and Biomedical Science

In order to define biomedical research, we need to first define biomedicine and biomedical science, terms that are frequently used, but not consistently well-defined. Biomedicine may be defined as the application of the principles of the natural sciences, especially biology and physiology, to clinical medicine [8]. Clinical medicine may be referred to as the study and practice of medicine based on direct observation of and interactions with patients. The term biomedical science frequently refers to the application of the principles of the natural sciences to medicine. There is a slight difference between biomedicine and biomedical science, with the former emphasizing the application of the principles of natural sciences to clinical medicine and the later embracing the application to the entire medicine, including both clinical and non-clinical medicine. Hence, biomedicine, biomedical science, and clinical medicine are tightly intertwined terms with significant overlaps and less obvious distinctions. This is particularly true as clinical medicine is becoming more and more scientific evidence-based.

3.2. Defining Biomedical Research

Similar to the difficulty encountered with defining biomedicine and biomedical science, the term biomedical research is also ill-defined though widely used in the literature and public media. As the term biomedical science refers to the application of the principles of natural sciences to medicine, the ultimate aim of biomedical research has been described as to answer medical questions leading to the discovery of treatment, prevention, and diagnosis of diseases that cause illness and death. In this context, biomedical research is also called medical research, and these two terms are often interchanged.

Biomedical research is a vast field of science that includes parts of life, physical, and social sciences. It is commonly divided into (i) basic biomedical research, (ii) clinical research, and (iii) translational research (**Figure 1-1**). This classification scheme has been widely used in both the United States and the Europe [9-14]. Indeed, clinical research (e.g., clinical trials) has been frequently included as a type of biomedical research in articles published in prestigious medical journals, including the Journal of American Medical Association (JAMA) and the New England Journal of Medicine (NEJM) (**Box 1-2**). However, the practice of including clinical research as part of biomedical research is not consistently adopted in the field of medical research. Indeed, many professionals at institutions of higher education view biomedical research and clinical research as two separate paths of medical research. In this regard, the terms biomedical research and basic biomedical research are frequently interchanged. Precisely, this is not correct.

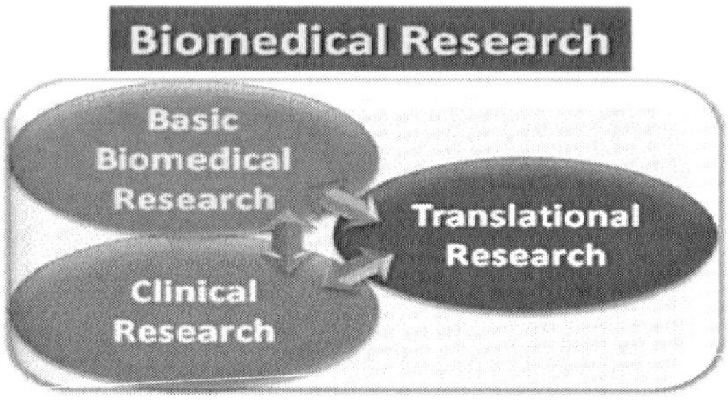

Figure 1-1. Scope of biomedical research. As illustrated, biomedical research comprises three intertwined components, i.e., basic biomedical research, translational research, and clinical research. Alternatively, biomedical research has been classified into two broad areas: biomedical research not involving human subjects, and biomedical research involving human subjects.

Box 1-2. JAMA and NEJM

JAMA: Journal of American Medical Association (www.jama.org)

- JAMA, published continuously since 1883, is an international peer-reviewed general medical journal published 48 times per year.
- JAMA is the most widely circulated medical journal in the world. The online version is made freely available to institutions in developing countries.
- JAMA's 2011 impact factor is 30 (the impact factor is a measure of citation rate per article, and is calculated by dividing one year's worth of citations to a journal's articles published in the previous two years by the number of major articles [e.g., research papers, reviews] published by that journal in those two years) (also see Section 3.4.5.1 and Box 1-7 for more discussion of impact factor).
- JAMA's acceptance rate is approximately 9% of the nearly 6000 solicited and unsolicited manuscripts it receives annually.
- The Editor-in-Chief of JAMA is Howard Bauchner, MD (as of July 2011) from Boston University School of Medicine and Public Health, MA, USA.

NEJM: New England Journal of Medicine (www.nejm.org)

- NEJM is the oldest continuously published medical periodical, completing its second century of service to the medical community in 2012. It is published 52 times per year.
- NEJM is dedicated to bringing physicians the best research and key information at the intersection of biomedical science and clinical practice, and to presenting the information in an understandable and clinically useful format.
- NEJM is the most widely read, cited, and influential general medical periodical in the world. Free online access is available in more than 100 low-income countries, and full text of research articles is free to all six months after publication.
- NEJM's 2011 impact factor is 53 (see JAMA above for how impact factor is determined).
- The Editor-in-Chief of NEJM is Jeffrey M. Drazen, MD (as of July 2000) from Harvard Medical School, MA, USA.

3.3. Defining Biomedical Research Scope

As noted above, biomedical research activities can be divided into three categories: (i) basic biomedical research, (ii) clinical research, and (iii) translational research. This section is intended to define the scope of each of the three categories.

3.3.1. Basic Biomedical Research

Basic biomedical research broadly investigates the underlying processes of living organisms to help understand how they function and malfunction at various levels including molecular, biochemical, and cellular levels. The World Health Organization (WHO) (**Box 1-3**) defines basic biomedical research as activities intended to find means of detecting, preventing, or treating human disease. Such research covers the discovery and exploratory studies that precede the regulated phases of drug development or programs to develop other methods of disease control [15]. Hence the WHO definition of basic biomedical research focuses on the basic research that leads to the development of modalities for the management of human diseases.

Box 1-3. The World Health Organization (WHO) (www.who.int)

History:

- When diplomats met to form the United Nations in 1945, one of the things they discussed was setting up a global health organization.
- WHO's Constitution came into force on April 7, 1948, a date we now celebrate every year as World Health Day.

Responsibilities:

- WHO is the directing and coordinating authority on international health within the United Nations' system.
- WHO experts produce health guidelines and standards, and help countries to address public health issues.
- WHO also supports and promotes health research. It shapes the research agenda and stimulates the generation, translation, and dissemination of valuable knowledge.
- Through WHO, governments can jointly tackle global health problems and improve people's well-being.

Director-General: Dr. Margaret Chan (as of November 2006) from China

Biomedical research is essential to the advancement of medicine and improvement of human health although basic biomedical research is receiving less public attention as reflected by its diminished presence in public policies [16-18]. Basic biomedical research frequently involves the use of isolated cells and tissue as well as live laboratory animals in the settings of wet laboratories with benches (see Chapter 5 for more detail). As such, basic biomedical research is frequently called laboratory bench work to distinguish it from clinical research.

3.3.2. Clinical Research

Clinical research can be defined broadly as investigation that applies basic research discoveries to human subjects to determine the effectiveness and safety of drugs, methods, and devices used to diagnose, support, and maintain individuals during and after treatment for diseases. In essence, clinical research is research that either directly involves individual people or uses materials from humans, such as their behavior or samples of their tissue that can be linked to a particular living person. The United States National Institutes of Health (NIH) (**Box 1-4**) defines clinical research in a comprehensive manner (**Table 1-1**) [19, 20].

Box 1-4. The United States National Institutes of Health (NIH) (www.nih.gov)

NIH at a glance:

- NIH, a part of the U.S. Department of Health and Human Services (HHS), is the nation's medical research agency-making important discoveries that improve health and save lives.
- NIH is the largest source of funding for medical research in the world, creating hundreds of thousands of high-quality jobs by funding thousands of scientists in universities and research institutions in every state across America and around the globe.
- NIH is made up of 27 institutes and centers, each with a specific research agenda, often focusing on particular diseases or body systems.

Mission of NIH:

- To seek fundamental knowledge about the nature and behavior of living systems and the application of that knowledge to enhance health, lengthen life, and reduce the burdens of illness and disability.

Goals of NIH:

- To foster fundamental creative discoveries, innovative research strategies, and their applications as a basis for ultimately protecting and improving health
- To develop, maintain, and renew scientific human and physical resources that will ensure the Nation's capability to prevent disease
- To expand the knowledge base in medical and associated sciences in order to enhance the Nation's economic well-being and ensure a continued high return on the public investment in research
- To exemplify and promote the highest level of scientific integrity, public accountability, and social responsibility in the conduct of science

Director: Francis S. Collins, MD, PhD (as of August 2009)

In 1997, the NIH Director's Panel on Clinical Research issued a three-part definition of clinical research (**Table 1-1**) [19, 20]. In 1998, the NIH introduced the Clinical Research Curriculum Award to improve the quality of training in clinical research. Over 50 training programs were funded through this award, and many of these programs grant degrees in clinical research [21]. The NIH's definition of clinical research has been widely accepted by institutions and programs and provides a common basis for the NIH-funded clinical research training programs. The definition has facilitated cross-program efforts to identify core competencies, best practices, and meaningful outcomes that are relevant across the broad spectrum of training in clinical research. This in turn has allowed program evaluators to develop useful assessment metrics to document the success of training programs. Today, about 30% of the NIH budget is spent on clinical research [21, 22]. To facilitate clinical trials, the NIH also provides a registry and results database of publicly and privately supported clinical studies of human participants conducted around the world. This database is known as ClinicalTrials.gov. ClinicalTrials.gov currently lists 147090 studies with locations in all 50 states of the United States and in 184 countries.

Table 1-1. NIH definition of clinical research involving three parts

SCOPE OF CLINICAL RESEARCH	DESCRIPTION
Patient-Oriented Research	This type of research involves a particular person or group of people or uses materials from humans. This research can include: • Studies of mechanisms of human disease • Studies of therapies or interventions for disease • Clinical trials • Studies to develop new technology related to disease
Epidemiological and Behavioral Studies	These types of studies examine: • The distribution of disease • The factors that affect health • How people make health-related decisions
Outcomes and Health Services Research	These studies seek to identify the most effective and most efficient • Interventions • Treatments • Services

As shown in **Table 1-1**, the term clinical trial is included in the patient-oriented research, and a clinical trial is defined by NIH as a prospective biomedical or behavioral research study of human subjects that is designed to answer specific questions about biomedical or behavioral interventions (such as drugs, treatments, devices, or new ways of using known drugs, treatments, or devices). The NIH definition of clinical trials is outlined below [23].

- Clinical trials are used to determine whether new biomedical or behavioral interventions are safe, efficacious, and effective.
- Behavioral human subjects research involving an intervention to modify behavior (such as diet, physical activity, cognitive therapy, etc.) fits this definition.
- Research with human subjects to develop or evaluate clinical laboratory tests (imaging or molecular diagnostic tests) might be considered as a clinical trial if the test will be used for medical decision making, or if the test itself imposes more than minimal risk for subjects.

Clinical trials used in drug development are sometimes described by phases. These phases are defined by the United States Food and Drug Administration (FDA) (**Box 1-5**). Each phase is designed to answer a separate research question (**Table 1-2**).

Box 1-5. The United States Food and Drug Administration (FDA) (www.fda.gov)

FDA at a glance:

- FDA is an agency within the U.S. Department of Health and Human Services.
- It consists of the Office of the Commissioner and four directorates overseeing the core functions of the agency: Medical Products and Tobacco, Foods, Global Regulatory Operations and Policy, and Operations.

Office of the Commissioner
Office of Foods
Office of Global Regulatory Operations and Policy
Office of Medical Products and Tobacco
Office of Operations

Responsibilities:

FDA's responsibilities extend to the 50 United States, the District of Columbia, Puerto Rico, Guam, the Virgin Islands, American Samoa, and other U.S. territories and possessions. FDA is responsible for:

- Protecting the public health by assuring that foods are safe, wholesome, sanitary and properly labeled; human and veterinary drugs, and vaccines and other biological products and medical devices intended for human use are safe and effective
- Protecting the public from electronic product radiation
- Assuring cosmetics and dietary supplements are safe and properly labeled
- Regulating tobacco products
- Advancing the public health by helping to speed product innovations
- Helping the public get accurate science-based information they need to use medicines, devices, and foods to improve their health

Commissiner: Margaret Hamburg, MD (as of May 2009).

Similarly, as pertaining to the review of applications submitted to NIH, clinical trials of experimental drug, treatment, device or behavioral intervention are described to proceed through four phases (**Table 1-3**).

It is worth mentioning here that a phase III clinical trial is specifically defined by NIH as a broadly based prospective clinical investigation, usually involving several hundred or more human subjects, for the purpose of evaluating an experimental intervention in comparison with a standard or controlled intervention or comparing two or more existing treatments [23]. Often the aim of such investigation is to provide evidence leading to a scientific basis for consideration of a change in health policy or standard of care. The definition includes pharmacological, non-pharmacological, and behavioral interventions given for disease prevention, prophylaxis, diagnosis, or therapy. Community trials and other population-based intervention trials are also included.

Table 1-2. Five phases of clinical trials in drug development as defined by FDA

PHASE OF CLINICAL TRIALS	DESCRIPTION
Phase 0	Exploratory study involving very limited human exposure to the drug, with no therapeutic or diagnostic goals (for example, screening studies, microdose studies).
Phase 1 (I)	Studies that are usually conducted with healthy volunteers and that emphasize safety. The goal is to find out what the drug's most frequent and serious adverse events are and, often, how the drug is metabolized and excreted.
Phase 2 (II)	Studies that gather preliminary data on effectiveness (whether the drug works in people who have a certain disease or condition). For example, participants receiving the drug may be compared with similar participants receiving a different treatment, usually an inactive substance (called a placebo) or a different drug. Safety continues to be evaluated, and short-term adverse events are studied.
Phase 3 (III)	Studies that gather more information about safety and effectiveness by studying different populations and different dosages and by using the drug in combination with other drugs.
Phase 4 (IV)	Studies that are carried out after the drug or treatment has been marketed to gather information on the drug's effect in various populations and any side effects associated with long-term use.

Table 1-3. Four phases of clinical trials described by NIH as pertaining to review of applications submitted to NIH

PHASE OF CLINICAL TRIAL	DESCRIPTION
Phase I	Phase I clinical trials are done to test a new biomedical intervention in a small group of people (e.g., 20-80) for the first time to evaluate safety (e.g., determine a safe dosage range, and identify side effects).
Phase II	Phase II clinical trials are done to study the biomedical or behavioral intervention in a larger group of people (several hundred) to determine efficacy and to further evaluate its safety.
Phase III	Phase III clinical trials are done to study the efficacy of the biomedical or behavioral intervention in large groups of human subjects (from several hundred to several thousand) by comparing the intervention to other standards or experimental interventions as well as to monitor adverse effects, and to collect information that will allow the intervention to be used safely.
Phase IV	Phase IV clinical trials are done after the intervention has been marketed. These studies are designed to monitor effectiveness of the approved intervention in the general population and to collect information about any adverse effects associated with widespread use.

3.3.3. Translational Research

Translational research has been a buzzword for many years now, and a Medline search indicated that the term translational research appeared as early as 1993 [21]. Translational research has been defined in various ways. Summarized below are the definitions from the NIH Clinical and Translational Science Award (CTSA) program, National Cancer Institute (NCI), and the Evaluation Committee of the Association for Clinical Research Training

(ACRT; now merged into ACTS; see the legend of **Figure 1-4** for more information on ACTS) [21]. ACTS stands for the Association for Clinical and Translational Science.

3.3.3.1. Defining Translational Research by NIH Clinical and Translational Science Award Program

One of the goals of the National Institutes of Health is to make translational research an integral part of biomedical research. In this context, NIH has made translational research a priority, forming centers of translational research at its institutes and launching the Clinical and Translational Science Award (CTSA) program in 2006. Impressively, the new National Center for Advancing Translational Sciences (NCATS), a signature project of NIH director Francis S. Collins, received congressional approval in the recently approved fiscal year 2012 spending package. NCATS was established in December 2011 and has been up and running since then with its mission being to catalyze innovations aimed at enhancing the development, testing, and implementation of diagnostics and therapeutics across a wide range of human diseases and conditions. CTSA has now been integrated into NCATS.

In an early announcement about applying for a CTSA before the establishment of NCATS, the NIH offered the following definition for translational research:

- Translational research includes two areas of translation.
- One is the process of applying discoveries generated during research in the laboratory, and in preclinical studies to the development of trials and studies in humans.
- The second area of translation concerns research aimed at enhancing the adoption of best practices in the community.
- Cost-effectiveness of prevention and treatment strategies is also an important part of translational science.

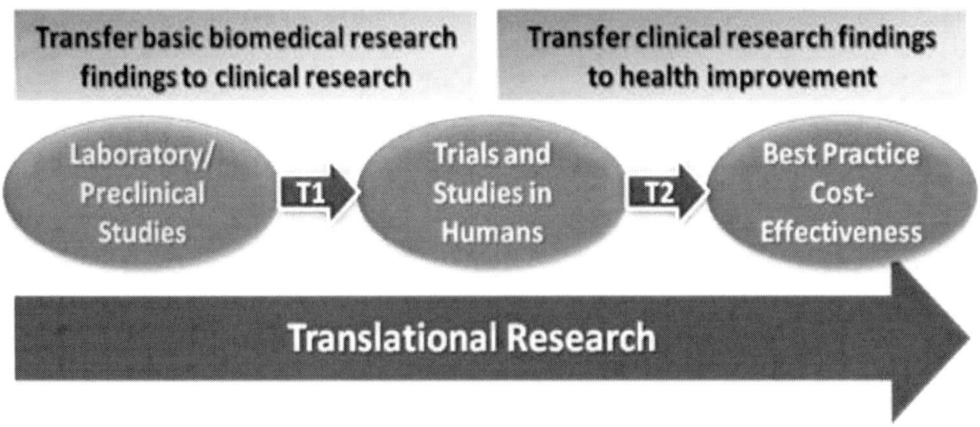

Figure 1-2. Schematic illustration of translational research, as defined by the NIH CTSA program before the establishment of NCATS. See text (Section 3.3.3.1) for detailed description. CTSA, Clinical and Translational Science Award; NCATS, National Center for Advancing Translational Sciences.

According to this definition, translational research is part of a unidirectional continuum in which research findings are moved from the researcher's bench to the patient's bedside and lastly to the community (**Figure 1-2**).

In the continuum, the first stage of translational research (T1) transfers knowledge from basic research to clinical research, while the second stage (T2) transfers findings from clinical studies or clinical trials to practice settings and communities, where the findings improve human health.

In a more recent program announcement of the Institutional Clinical and Translational Science Award (U54), the newly established NCATS defines translational research broadly to include the early steps necessary to develop new therapeutics, devices and diagnostics from basic discoveries, the steps necessary to establish real world efficacy, and the research needed to improve the practical implementation and dissemination of improved approaches to care. This breadth is sometimes described as T1 through T4 (**Table 1-4**) [24]. The NCATS definition of translational research is becoming widely accepted as the most comprehensive description of translational research.

Table 1-4. NCATS definition of translational research from T1 to T4

PHASE OF TRANSLATION	DESCRIPTION
T1: Translation to Humans	Testing basic science discoveries for clinical effect and/or applicability
T2: Translation to Patients	Testing new interventions in human subjects under controlled environments to form the basis for clinical applications and evidence-based guidelines
T3: Translation to Practice	Research on the application of new interventions or therapies in general practice; research that yields knowledge on best ways to implement new medical interventions in the clinic
T4: Translation to Population	Investigations of factors and/or interventions that influence the health of populations, which ultimately results in improved health of the public

3.3.3.2. DEFINING TRANSLATIONAL RESEARCH BY NCI

In an effort to develop a framework for translational research on cancer, the Translational Research Working Group (TRWG) of the National Institute of Cancer (NCI) defines translational research in the following way (**Figure 1-3**):

"Translational research transforms scientific discoveries arising from laboratory, clinical, or population studies into clinical applications to reduce cancer incidence, morbidity, and mortality [25]." As noted, this definition is specific for research on cancer, emphasizing the impact of translational research on the incidence, morbidity, and mortality of cancer, which is the number two cause of death in the United States and also a major cause of mortality worldwide.

3.3.3.3. DEFINING TRANSLATIONAL RESEARCH BY THE ASSOCIATION FOR CLINICAL RESEARCH TRAINING

Recently, the Evaluation Committee of the Association for Clinical Research Training (ACRT) has developed a working definition of translational research, which is described below and also illustrated in **Figure 1-4** [21]. ACRT has recently been merged into a single new organization- the Association for Clinical and Translational Science (ACTS; also see the legend of **Figure 1-4**).

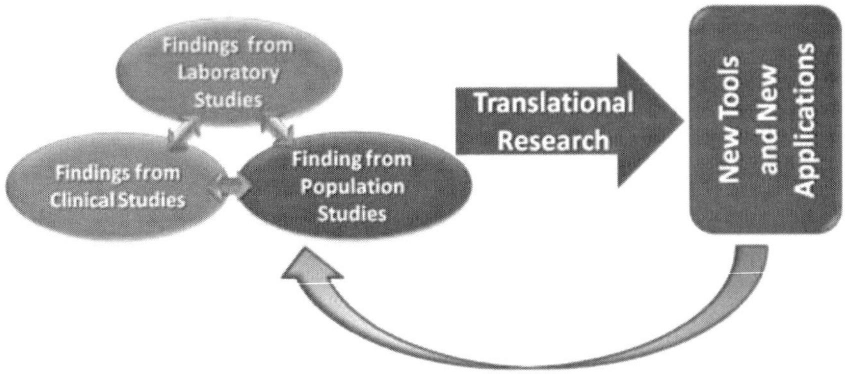

Figure 1-3. Translational research as defined by NCI TRWG. NCI, National Cancer Institute; TRWG, Translational Research Working Group.

As illustrated in **Figure 1-4**, translational research fosters the multidirectional integration of basic research, patient-oriented research, and population-based research, with the long-term aim of improving the health of the public. The entire process involves three types of translation, T1, T2, and T3, as outlined below.

- T1 research expedites the movement between basic research and patient-oriented research that leads to new or improved scientific understanding or standards of care.
- T2 research facilitates the movement between patient-oriented research and population-based research that leads to better patient outcomes, the implementation of best practices, and improved health status in communities.
- T3 research promotes interaction between laboratory-based research and population-based research to stimulate a robust scientific understanding of human health and disease.

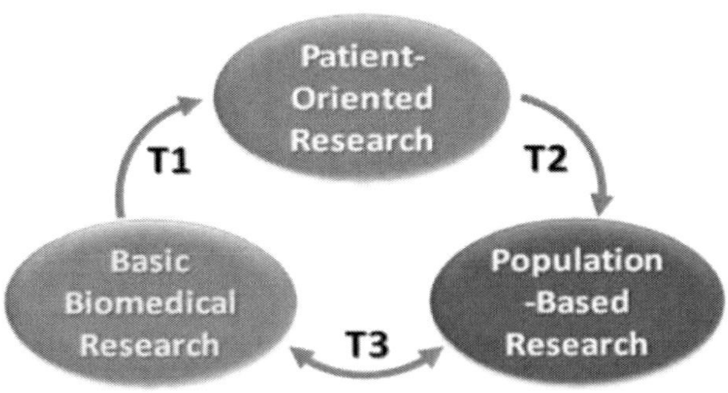

Figure 1-4. Translational research as defined by the Evaluation Committee of the ACRT [21]. This definition emphasizes that translational research moves in a bidirectional manner from one type of research to another, from basic research to patient-oriented research, to population-based research, and back, and involves collaboration among scientists from multiple disciplines. In addition to ACRT, many other organizations support clinical and translational research and the training of the next generation of translational research scientists in the United States as well as globally. These include the Association for Patient Oriented Research (APOR) and the Society for Clinical and Translational Science (SCTS). To better meet the goal of supporting the translational research and training, ACRT, APOR, and SCTS have recently been merged into a single new organization, the Association for Clinical and Translational Science (ACTS; http://www.ctssociety.org).

As translational science advances, its contribution to the optimization of clinical practices and the improvement of human health is becoming increasingly evident. This is not only true for translation research, but also for biomedical research in general [13, 26, 27]. In fact, the impact of biomedical research goes beyond the scope of human health and moves into the domain of social and economic development [28] (see Chapter 2 for detailed discussion). While we are enjoying the benefits of biomedical research, it is important to note that the continued success of biomedical research relies on the strict adherence to the highest standards of responsible conduct of biomedical research. Section 4 discusses the general principles of biomedical research ethics and responsible conduct of biomedical research.

3.4. Defining Biomedical Research Process

Biomedical research is the process of identifying scientific questions, choosing, and applying the most suitable scientific methods in the context of a rigorous research design to answer the questions, and finally disseminating the findings to contribute to the knowledge base and benefit human health. As illustrated in **Figure 1-5**, biomedical research process follows the major steps of conducting scientific research as outlined earlier in Section 2, and includes: (1) identifying the research questions and searching and reviewing the literature, (2) formulating the research hypothesis and developing the specific aims, (3) choosing the appropriate methodology and formulating a rigorous research design for collecting and analyzing research data, (4) implementing research and collecting, managing, and analyzing the research data, and (5) reporting and disseminating the research findings.

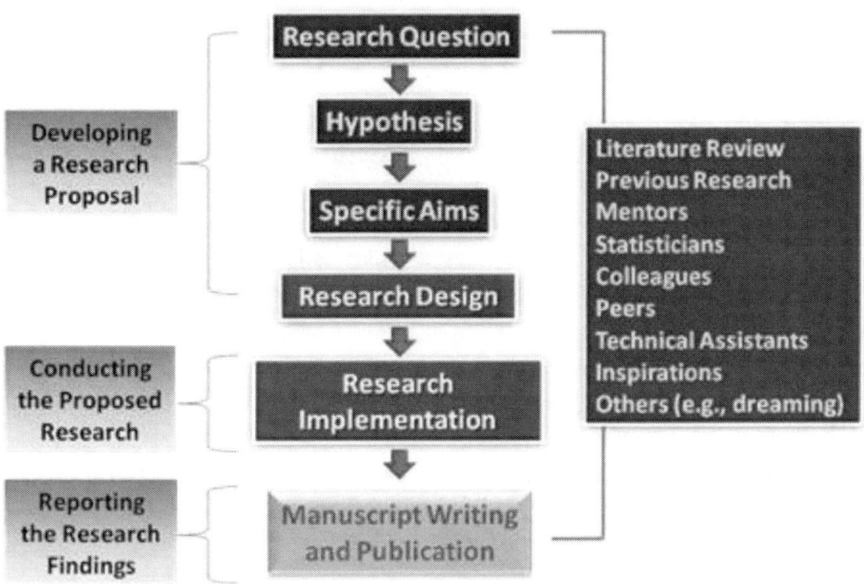

Figure 1-5. Schematic illustration of the biomedical research process. As depicted, biomedical research process consists of three phases: (i) developing a research proposal, (ii) conducting the proposed research, and (iii) reporting the research findings. Each phase includes multiple steps. See text (Section 3.4) for detailed description of the individual steps of biomedical research process.

3.4.1. Identifying the Research Questions and Searching and Reviewing the Literature

3.4.1.1. DEFINING RESEARCH QUESTIONS

A research question may be defined as a statement that identifies the phenomenon to be studied. Biomedical research questions may originate from a number of different sources, including previous research conducted by the same investigators or others, the findings reported in the literature, and the investigators' personal observations (such as an observation by a physician on an unusual drug response in his/her patients). Among the above sources, searching and reviewing the literature is becoming more and more important in identifying biomedical research questions as well as in formulating a sound hypothesis. An effective way of identifying relevant literature is through searching the electronic databases. Among the various electronic sources, MEDLINE (**Box 1-6**) is probably the most extensively used database for searching biomedical literature.

Box 1-6. MEDLINE at a glance
(http://www.nlm.nih.gov/pubs/factsheets/medline.html)

- MEDLINE is the United States National Library of Medicine's (NLM) premier bibliographic database that contains over 19 million references to journal articles in life sciences with a concentration on biomedicine.
- MEDLINE is the primary component of PubMed, part of the Entrez series of databases provided by the NLM National Center for Biotechnology Information (NCBI). Searching is free of charge and does not require registration.
- Time coverage of MEDLINE is generally from 1946 to the present, with some older material.
- Currently, citations of MEDLINE are from approximately 5,600 worldwide journals in 39 languages.
- The subject scope of MEDLINE is biomedicine and health, broadly defined to encompass those areas of the life sciences, behavioral sciences, chemical sciences, and bioengineering needed by health professionals and others engaged in basic research and clinical care, public health, health policy development, or related educational activities. MEDLINE also covers life sciences vital to biomedical practitioners, researchers, and educators, including aspects of biology, environmental science, marine biology, plant and animal science, as well as biophysics and chemistry.
- The majority of the publications covered in MEDLINE are scholarly journals; a small number of newspapers, magazines, and newsletters considered useful to particular segments of the NLM broad user community are also included.

3.4.1.2. CLASSIFYING RESEARCH QUESTIONS

Biomedical research questions are broadly classified into the following three categories: (1) descriptive questions that ask what is going on or what exists (e.g., what is the prevalence of obesity in children in a particular city? What is the lethal dose of a new drug in laboratory mice?); (2) relational questions that ask the relationship between two or more variables (e.g., what is the relationship between height and cancer risk? What is the relationship between the antioxidant activity of nutraceuticals and their protective effects against chemical carcinogenesis in laboratory rats?); and (3) causal questions that ask if there is a causal linkage between two or more variables (e.g., does daily intake of a low dose of aspirin decrease the incidence of prostatic cancer in high risk individuals? Does depletion of glutathione in liver via tissue-specific gene inactivation cause hepatic inflammation in rats?).

Table 1-5. The three fundamental types of causal relationships [29]

Causal Relationship	Description
Sufficient Cause	If the factor (cause) is present, the effect (disease) will always occur (e.g., homozygous hexosaminidase A deficiency inevitably leads to Tay-Sachs disease).
Necessary Cause	The factor (cause) must be present for the effect (disease) to occur. A necessary cause may be present without the disease occurring (e.g., without the presence of *M. tuberculosis*, the disease tuberculosis cannot occur. However, people may have the *M. tuberculosis* bacteria in their bodies but do not develop tuberculosis).
Risk Factor	A risk factor is an exposure, behavior, or attribute that, if present and active, clearly increases the probability of a particular disease in a group of people who have the risk factor compared with an otherwise similar group of people who do not (e.g., cigarette smoking is a risk factor for lung cancer; hypercholesterolemia is a risk factor for coronary artery disease).

Table 1-6. Major guidelines for judging whether an association is causal [30]

Guideline	Description
Temporal Relationship	If a factor is the cause of a disease, exposure to the factor must occur before disease develops (e.g., for *M. tuberculosis* to be the cause of tuberculosis, exposure to the bacteria must occur before the development of the disease).
Strength of the Association	The strength of an association is measured by the relative risk (or odds ratio). The stronger the association, the most likely it is that the relation is causal. Here the term odds is defined as the ratio of the probability of occurrence of an event to that of non-occurrence, or the ratio of the probability that something is so, to the probability that it is not so. The term odds ratio is defined as the ratio of the odds of an event in one group to the odds of an event in another group. An odds ratio of 1.0 indicates no difference between comparison groups. For undesirable outcomes an odds ratio that is <1.0 indicates that the intervention was effective in reducing the risk of that outcome.
Dose-Response Relationship	Dose-response relationship describes the change in effect on, or response of, an organism caused by differing levels of exposure (or doses) to a factor (usually a chemical, a drug, or a pathogen) after a certain exposure time. If a dose-response relationship is present, it is a strong evidence for a causal relationship. However, the absence of a dose-response relationship does not rule out a causal relationship.
Replication of the Findings	If the relationship is causal, it would be expected to be found consistently in different studies and in different populations.
Cessation of Exposure	If a factor is a cause of a disease, it would be expected that the risk or incidence of the disease declines as the exposure to the risk factor is reduced or eliminated.
Biological Plausibility	Biological plausibility refers to coherence with the current body of biological knowledge. A causal relationship between two variables would be explained by or consistent with existing biological knowledge.

Most scientific research seeks to identify causal relationships. Indeed, investigation of the causal relationships is one of the most important and challenging tasks of both basic and clinical biomedical research. **Table 1-5** lists the three fundamental types of causal relationships. The main guidelines for judging whether an association is causal are summarized in **Table 1-6**.

3.4.1.3. DEVELOPING GREAT RESEARCH QUESTIONS

Researchers spend a great deal of time in refining research ideas into testable research questions. To develop a strong biomedical research question from your idea, you should ask yourself the following four fundamental questions (**Table 1-7**): (1) do I know the field? (2) is the research question novel? (3) is the research question important? and (4) is the research question ethically testable?

Table 1-7. Questions that need to be asked in developing a great biomedical research question

QUESTION	DESCRIPTION
Do I Know the Field?	• Do I know the specific biomedical research field and the related literature well? • What are the important research questions in the specific biomedical research field that I am interested in? • What areas need further exploration in the particular research field?
Is the Research Question Novel?	• Has a great deal of research already been conducted in the topic area(s) that I am interested in? • Has this study been done before? If so, is there any room for improvement? • Could my study fill a gap and lead to a greater understanding?
Is the Research Question Important?	• Is it an important question that will have significant impact on the field? • Is the timing right for this question to be answered? • Is it a hot topic or is it becoming obsolete?
Is the Research Question Ethically Testable? Are there enough funds and other resources to support the proposed research?	• Is the conduction of the research ethical? • What are the guidelines and policies that I need to follow to ensure the ethical conduction of the research? • Does the research team have adequate expertise to carry out this research? • Can enough subjects be enrolled (for human subjects) or obtained (for laboratory animals)? • Are the required facilities and resources available to this research? • Is there any inhouse fund to support the research? • Is the research fundable by federal agencies, especially the United States National Institutes of Health? • Can the research question lead directly into hypothesis and be converted into specific aims?

3.4.2. Formulating the Research Hypothesis and Developing Specific Aims

A well-thought out and focused biomedical research question leads directly to the research hypothesis. A research hypothesis is the statement created by the researchers when they speculate upon the outcome of a biomedical research. Hypotheses are derived from the research questions and are more specific predictions of the relationship or differences between two or more variables in terms of expected results or outcomes of a study. Hypotheses are formulated before the study is actually conducted, and they provide direction for the development of the specific aims.

Along with the research question and hypothesis, well-defined specific aims are key to the successful completion of a biomedical research project. The specific aims should be hypothesis-driven and answer the question "What are you going to do to test your hypothesis?" It is common for a study to have between two and five specific aims that are components of an overall research question. An individual specific aim may also consist of several sub-aims depending on the nature of the proposed research. The National Institute of Allergy and Infectious Diseases (NIAID; an institute of the National Institutes of Health) provides very helpful guidance on writing specific aims [31].

3.4.3. Choosing the Appropriate Methodology and Formulating a Rigorous Research Design for Collecting and Analyzing the Research Data

Choosing the appropriate methodology and formulating a rigorous research design are the most challenging aspects of a research proposal. The term methodology may be broadly defined as a body of practices, procedures, and rules used by those who work in a discipline or engage in a scientific inquiry. Research design (also known as study design) may be defined as a blueprint that outlines the logistics regarding (1) the number and grouping of the research subjects (e.g., animals, humans) to be included, (2) variables and endpoints to be determined for testing the hypothesis, (3) the methods and procedures to be used for data recording, collection, and management, and finally (4) the data analysis and statistics. This section introduces the general principles of study design for basic biomedical research involving laboratory animals and for clinical research involving human subjects.

3.4.3.1. Study Design for Basic Biomedical Research

Table 1-8 summarizes the key components of a study design for basic biomedical research involving the use of laboratory animals. It should be noted that the study design is governed by the research hypothesis and specific aims. It is also important to mention that the regulatory policies and procedures related to animal care and use must be followed (see Chapters 4 and 7 for more details on responsible conduct of research involving the use of laboratory animals).

3.4.3.2. Study Design for Clinical Research

Over the past decade, the interest in clinical research has increased with emphasis placed on evidence-based, data-driven health care advancements. The validity of the clinical research findings can be threatened by issues of bias and confounding. Biases are systematic errors in how study subjects are selected or measured, which result in false inferences. Confounding is a distortion in findings that is attributable to mixing variable effects. Hence, understanding the principles of clinical study design is not only important to clinical research scientists, but also imperative for health care providers. As illustrated in **Figure 1-6**, there are three categories of clinical study design: (1) experimental clinical research design, (2) observational clinical research design, and (3) case studies or case report.

Table 1-8. Key components of a study design for basic biomedical research involving the use of laboratory animals

STUDY DESIGN COMPONENT	DESCRIPTION
Research Subjects (Animals)	• Animal species, animal models, sex, age, body weight, diet, and housing condition • Experimental groups and number of animals for each group (power-analysis for determining sample size) • Randomization of animals into experimental groups to control bias and unknown sources of variation • Positive and/or negative control groups if needed
Variables and Endpoints	• The variables to be investigated (e.g., changes of blood pressure in response to two different drugs; changes of inflammatory cytokines following treatment with a nutraceutical) • The specific endpoints of measurement (e.g., systolic blood pressure, diastolic blood pressure, plasma levels of interleukin-1 and tumor necrosis factor)
Methods, Procedures, Materials, and Instruments	• Established methods • Standardized experimental procedures • Materials (chemicals and reagents) of high purity and quality • Reliable instruments and standardized protocols for operating the instruments, including calibration
Data Recording, Collection, Management, and Analysis	• Standardized protocols for data recording and collection • Most suitable statistical analyses • Consultation with biostatisticians and other experts to formulate data management and analysis plans • The data management and analysis plans are critical components of the research design. • The data management plan should address the selections of database platforms, data elements to be collected, data entry, processes for ensuring data quality, and data security. • The investigator team must determine how the data will be analyzed and whether descriptive statistics, inferential statistics, or both will be used. • A biostatistician should be consulted early in the design phase. • Many new investigators make the fatal error of consulting with the biostatistician when data collection is complete and analysis is required. This can result in method flaws or inadequate data collection and can prevent useful analysis of the data [32].

Observational designs include the cohort study, case-control study, and cross-sectional study, while experimental clinical research design typically involves a randomized controlled trial (RCT). If the RCT is carried out in a blinded manner, it is called blinded randomized controlled trial. Case studies can be informative in certain situations, but have limitations due to the low strength of evidence that they provide. In the hierarchy of evidence provided by the various clinical study design modalities, the blinded randomized controlled trial provides the strongest evidence for the safety and efficacy of a treatment modality.

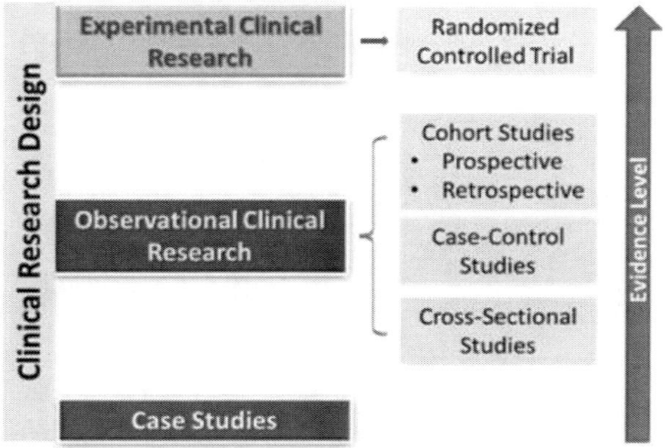

Figure 1-6. Clinical research design modalities and the level of evidence they provide.

The basic design of a randomized controlled trial is depicted in **Figure 1-7**, and the major characteristics of a blinded RCT are summarized below [29, 30, 33]. The definitions for cohort, case-control, and cross-sectional studies as well as some related terms are given in **Table 1-9**.

- Random assignment of intervention, or treatment, in which a group of patients are randomly assigned either to an experimental group to receive a treatment such as a new drug, or to a control group (the control group might receive nothing, placebo, or an active alternative).
- Randomization eliminates the influence of confounding variables that are present at the time of randomization. It does not eliminate the confounding variables that develop during the follow up, such as "cointervention". In an unblinded study (see next paragraph for discussion of blinding), the investigators give extra attention or treatment to participants that the investigators know are receiving the active treatment, and this "cointervention" may be the actual cause of any difference in outcome that is observed between the groups. This "cointervention" confounding variable can be eliminated by blinding.
- Whenever possible, the investigators should design the trial in such a way that the study participants, study staff who have contact with them, persons making laboratory measurements, and those adjudicating outcomes have no knowledge of the study group assignment. When it is not possible to blind all of these individuals, it is highly desirable to blind as many as possible and always blind laboratory personnel.
- Blinding includes single- and double-blind masking. Single-blind masking is a type of masking in which one party involved with the clinical trial, either the investigators or participants, does not know which participants have been assigned which interventions. Double-blind masking is a type of masking in which two or more parties involved with the clinical trial do not know which participants have been assigned with which interventions. Typically, this includes the investigators and participants. On the other hand, the term open-label study describes a clinical trial in which masking is not used. This means that all parties involved with the trial know which participants have been assigned with which interventions.

- Double-blind (also called double-blinded) randomized controlled trial offers the most solid basis for an inference of cause and effect, compared with the results obtained from any other clinical study design modalities.

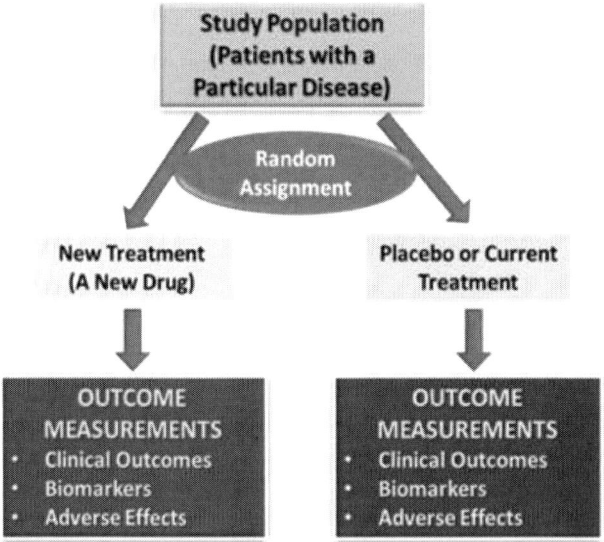

Figure 1-7. Basic design of a randomized controlled trial. A placebo is a substance or other kind of treatment that looks just like a regular treatment or medicine, but it's not. It's actually an inactive "look-alike" treatment or substance. The person getting a placebo does not know for sure that the treatment is not real. Sometimes the placebo is in the form of a "sugar pill," but a placebo can also be an injection, a liquid, or even a procedure. Even though they do not act on the disease, placebos seem to affect how people feel (this happens in up to 1 out of 3 patients). A change in a person's symptoms as a result of getting a placebo is called the placebo effect.

Table 1-9. Definitions of cohort, case-control, and cross-sectional studies and some related terms

TERM	DEFINITION
Cohort Study	• A cohort study is a type of observational study in which the investigators select a group of exposed individuals and a group of non-exposed individuals and then follow up both groups to compare the incidence of disease. • The cohort design may include more than two groups. • Cohort studies can be either prospective or retrospective. • In a prospective cohort study, the cohort (exposed group and non-exposed group) is formed and then followed over time. • In a retrospective cohort study, data is gathered for a cohort that was formed sometime in the past where good records have been kept and all exposures and outcomes have already occurred. • In both retrospective and prospective cohort studies, the goal is to break down the cohort into a group of exposed and a group of non-exposed individuals and follow their outcomes (such as incidence of a disease) over time to investigate the relationship between the exposure and the incidence of the disease.

TERM	DEFINITION
Case-Control Study	• A case-control study is also a type of observational study that compares two groups of people: those with the disease or condition under study (cases) and a very similar group of people who do not have the disease or condition (controls). • Investigators study the medical and lifestyle histories of the people in each group to learn what factors may be associated with the disease or condition. For example, one group may have been exposed to a particular substance that the other was not.
Cross-Sectional Study	• A cross sectional study measures the prevalence of health outcomes or determinants of health, or both, in a population at a point in time or over a short period. • Such information can be used to explore disease etiology. • For example, the relation between cataract and vitamin status has been examined in cross sectional surveys. • However, associations must be interpreted with caution. Bias may arise because of selection into or out of the study population.
Case Study/Report	• A case study or case report describes a single or a few of cases of a disease, usually with an unexpected presentation. • A case report typically describes the findings, clinical course, and prognosis of one or a few cases, often accompanied by a review of other cases previously reported in the biomedical literature. • A case report may also be on a positional or quantitative variation of the anatomical structures. • A case report provides a type of anecdotal evidence. As such, it is less scientifically rigorous than data from randomized controlled clinical trials involving a larger sample size. Nevertheless, a case report may shed new light on the possible cause or pathogenesis of a disease, or a novel therapeutic or adverse effect of a drug.
Systematic Review	• A systematic review is a summary of the clinical literature in a comprehensive manner. It is a critical assessment and evaluation of all available clinical studies that address a particular clinical issue. • The researchers use an organized method of locating, assembling, and evaluating a body of literature on a particular topic using a set of specific criteria. • A systematic review typically includes a description of the findings of the collection of research studies. • For example, scientists collected all the published studies that compared types of treatment for prostate cancer that had not spread beyond the prostate gland. They compiled the results of these studies in a comparative effectiveness review, which is a type of systematic review.
Meta-Analysis	• Meta-analysis may be defined as a way of combining data from many different research studies. A meta-analysis is a statistical process that combines the findings from individual studies. • For example, researchers wanted to know about the risk of stomach bleeding in people taking aspirin. They did a meta-analysis of data from 24 clinical trials with nearly 66,000 participants and found that the risk of stomach bleeding was 2.47 percent with aspirin compared to 1.42 percent with placebo (inactive substance). • The systematic review may also include a quantitative pooling of data, known as a meta-analysis. Hence, the compound term "systematic review and meta-analysis" in commonly encountered in the biomedical literature. • Meta-analysis of randomized controlled trials provides much stronger evidence than the individual trials.

Table 1-9. (Continued)

TERM	DEFINITION
Evidence-Based Practice	• Evidence-based practice means that applying the best available research results (evidence) when making decisions about health care. • Health care professionals who perform evidence-based practice use research evidence along with clinical expertise and patient preferences. Systematic reviews and meta-analysis provide information that aids in the process of evidence-based practice. • For example, a health care provider recommends acetaminophen to treat arthritis pain in a patient who has recently had stomach bleeding. The health care provider makes this recommendation because research shows that acetaminophen is associated with less risk for stomach bleeds than other common pain relievers. The health care provider's recommendation is an example of evidenced-based practice.

3.4.4. Implementing Research and Collecting, Managing, and Analyzing the Research Data

Good Laboratory Practice (GLP) and Good Clinical Practice (GCP) (**Table 1-10**) are required by regulatory agencies, such as the United States Food and Drug Administration (U.S. FDA) for conducting preclinical and clinical research, respectively, for evaluating the safety and efficacy of pharmaceuticals and for assessing the safety of other substances. Even for studies that are not regulated by federal agencies, it is prudent to get into the habit of following GLP for basic biomedical research and GCP for studies involving human subjects. The internationally-recognized detailed guidelines for GLP and GCP are available online for free [34, 35].

Table 1-10. Definitions of Good Laboratory Practice and Good Clinical Practice

TERM	DEFINITION
Good Laboratory Practice (GLP)	• GLP applies to non-clinical studies conducted for the assessment of the safety or efficacy of chemicals (including pharmaceuticals) to humans, animals and the environment. • An internationally recognized definition for GLP is that GLP embodies a set of principles that provides a framework within which studies are planned, performed, monitored, recorded, reported and archived. These studies are undertaken to generate data by which the hazards and risks to users, consumers and third parties, including the environment, can be assessed for pharmaceuticals, agrochemicals, veterinary medicines, industrial chemicals, cosmetics, food and feed additives and biocides. • GLP helps assure regulatory authorities that the data submitted are a true reflection of the results obtained during the study and can therefore be relied upon when making risk/safety assessments. • The U.S. FDA Code of Federal Regulations 21CFR58: Good Laboratory Practice for Preclinical Studies [34] prescribes good laboratory practices for conducting nonclinical laboratory studies that support or are intended to support applications for research or marketing permits for products regulated by the FDA, including food and color additives, animal food additives, human and animal drugs, medical devices for human use, biological products, and electronic products.

TERM	DEFINITION
Good Clinical Practice (GCP)	• GCP is an international ethical and scientific quality standard for designing, conducting, recording, and reporting trials that involve the participation of human subjects. • Compliance with GCP standard provides public assurance that the rights, safety, and wellbeing of trial subjects are protected, consistent with the principles that have their origin in the Declaration of Helsinki, and that the clinical trial data are credible (see Chapter 8). • Adherence to the principles of GCP is universally recognized as a critical requirement to the conduct of research involving human subjects. • The International Conference on Harmonisation (ICH) GCP guidance (1996) [35] is the internationally recognized guideline that provides a unified standard for the European Union (EU), Japan, and the United States to facilitate the mutual acceptance of clinical data by the regulatory authorities in these jurisdictions.

3.4.5. Reporting and Disseminating the Research Findings

Plans for presentation and publication of the research findings should be considered before completing the process of data collection and analysis. The findings can be presented in a scientific meeting in the format of either a poster or an oral presentation. However, publication in a peer-reviewed biomedical journal carries more weight and facilitates the quick dissemination of the research findings. Many new investigators and even experienced researchers become overwhelmed by the process of writing and submitting a manuscript for publication in a biomedical journal. Indeed, writing a good scientific manuscript is not easy. Many of the new investigators would rather think of new experiments, carry out experiments, analyze the results of experiments, or even draw figures from the results than write the manuscript. However, there isn't much point in doing the experiments if the work is never published and disseminated. Hence, learning how to write good scientific manuscripts is an indispensible process in a person's scientific career.

Describing tips for or detailed practical approaches to writing scientific manuscripts is beyond the scope of this chapter. This section is intended to discuss the key steps involved in writing and submitting scientific manuscripts for publication in biomedical journals. As illustrated in **Figure 1-8**, publication of a scientific manuscript in a biomedical journal involves multiple key steps: (1) identifying a suitable journal, (2) writing the manuscript based on the format requirements of the journal, (3) submitting the manuscript to the journal for peer-review, and (4) revising the manuscript according to the reviewers' comments and resubmitting it to the journal.

3.4.5.1. IDENTIFYING A SUITABLE BIOMEDICAL JOURNAL

Researchers are usually passionate about their research findings and wish to have their manuscripts published in high-tier biomedical journals. A commonly used way of ranking biomedical journals is by the impact factors (**Box 1-7**). However, impact factor is not a perfect way for assessing the quality of a journal. In this context, many subspecialty journals do not have very high impact factors, yet are highly respected in the field. Likewise, quality and value of a research manuscript cannot solely be judged by the journal where it is published. Many important findings are in fact published in subspecialty journals with moderate impact factors. It is the quality of the work, not the journal where it is published that determines the value and impact of the manuscript.

Figure 1-8. Major steps involved in publication of a biomedical research manuscript.

Box 1-7. Journal impact factor: definition, calculation, and caveats

Definition:

- Journal impact factor is a measure of the frequency with which the "average article" in a journal has been cited in a given period of time.
- Journal impact factor is released from Journal Citation Report (JCR), a product of Thomson Institute for Scientific Information (ISI), now part of Thomson Reuters. The JCR provides quantitative tools for ranking, evaluating, categorizing, and comparing journals. The impact factor is one of these.
- The annual JCR impact factor is a ratio between citations and recent citable items published.

Calculation:

- The impact factor of a journal is calculated by dividing the number of current year citations to the source items published in that journal during the previous two years. For example, the 2012 impact factor for Journal S would be calculated as follows:

  A = the number of times that articles published in 2010-2011 in Journal S were cited in indexed journals during 2012

B = the total number of articles published in 2010- 2011 in Journal S

The 2012 impact factor for Journal S = A/B

- If, for example, a total of 100 articles were published in Journal S in 2010-2011 (two years), and the total citations in 2012 to the articles published in Journal S in 2010-2011 were 300 (this could be that only one of the 100 articles published in 2010-2011 was cited in 2012, but it was cited 300 times; or could be that all the 100 articles were cited in 2012, but once for each article), the 2012 impact factor for Journal S is 300/100 = 3.

Note: the 2012 impact factor is published in 2013, because it could not be calculated until all of the 2012 publications had been received. The 2013 impact factor will be published in 2014.

Caveats:

- Thomson Reuters does not depend on the impact factor alone in assessing the usefulness of a journal, and neither should anyone else.
- The impact factor should not be used without careful attention to the many phenomena that influence citation rates, as for example the average number of references cited in the average article in a journal.
- There are also many artifacts that can influence a journal's impact factor and its ranking in journal lists, not the least of which is the inclusion of review articles or letters.
- Review articles generally are cited more frequently than typical research articles because they often serve as surrogates for earlier literature, especially in journals that discourage extensive bibliographies.
- Journals that only publish review articles usually have much higher impact factors than do regular journals.
- General journals tend to have higher impact factors than subspecialty journals. However, subspecialty journals are more likely to have a more focused audience. In this context, many well-respected specialty journals have only moderate impact factors.

3.4.5.2. Writing the Manuscript based on the Format Requirements of the Journal

Different biomedical journals have different format and submission requirements. It is suggested that the researchers carefully review the desired journal's author instructions in order to understand the format and submission requirements.

Although it depends on the standards of the journal to which you choose to submit your manuscript, a general format you can follow for writing your manuscript may include the following sections (**Table 1-11**): (1) title/face page including title of the manuscript and authors and their affiliations, among others (2) abstract, (3) introduction, (4) materials and methods, (5) results, (6) discussion and conclusion, (7) acknowledgements, (8) references, (9) tables and illustrations, and (10) supplementary materials.

Table 1-11. Structural components of a typical biomedical research manuscript

SECTION	DESCRIPTION
Title/Face Page	• Descriptive title • Short-running title: a mini-version of the title • Authors and affiliations • Corresponding author and contact information: Corresponding author is the only author who has the right to submit the manuscript and communicate with the journal's editorial office or editor regarding the publication of the manuscript. Many journals allow more than one corresponding author for a manuscript. • Others, such as a list of abbreviations used and notes
Abstract and Key Words	• Abstract is a crucial part of a scientific article. • Abstract along with title and authorship/affiliations are the only components that are made available by journals free to the readers on their websites and through the electronic databases such as MEDLINE (Box 1-6). • Abstract is considered a mini-version of the manuscript and should contain information on background, purpose, aims of the study, key data, and major conclusions. • Abstract is usually written in ~150-300 words depending on the journals. • Usually 3-6 key words are required by most biomedical journals.
Introduction	• Introduction is the first section of the main text. • It should provide brief background information and focus on justifying why the research being reported was undertaken. • Introduction section can be from 250 to 750 words depending on the nature of the research project.
Materials and Methods	• This section describes how the research being reported was carried out. • The materials part should list the major materials used and their sources. • The methods and procedures part should provide sufficient detail for the reader to be able to reproduce your study. • Statistical analysis is included in this section. • The Materials and Methods section can be from 250 to 1000 words, and some journals publish the detailed methods and procedures as supplementary material.
Results	• The Results section describes all the key data and findings in a concise and clear manner. • The data can be presented in tables and illustrations (e.g., figures, charts, photographs). • Please note that most journals ask the authors not to insert the tables/illustrations into the text. They should be uploaded as separate files during online submission, and usually placed at the end of the assembled full manuscript after the References section. • The length of the Results section varies and depends on the amount of data to be presented. The length of the text may range from 250 to 750 words.
Discussion and Conclusion	• The Discussion section describes the significance of your findings. • It should not only discuss the strengths, but also the weaknesses and limitations of your work. • This is the place where you explain or interpret your data in the context of current knowledge base and formulate your own theory or new hypothesis. • You may also discuss the clinical implications of your findings and point to the future directions. • At the end of the Discussion section, you may use 3-5 sentences to summarize the principal findings of your research and their implications. Alternatively, this could be included in a separate Conclusion section after the Discussion.

SECTION	DESCRIPTION
	• The Discussion section is usually written in 250-750 words. Lengthy discussion should be avoided.
Acknowledgements, Declaration of Interest, and Others	• These may be listed after the References section in some journals. • Those who contributed to the work but are not listed as authors should be acknowledged. • The funding source(s) should be acknowledged. • Some journals require that the responsibilities of the authors in the work be briefly described. • Declaration of interest is an important section to complete especially for authors involved in clinical trials.
References	• Biomedical science moves forward by building upon the work of others. • It is important to appropriately cite references in your scientific manuscripts to support your statements or to indicate the sources of information, quotations, or borrowed ideas presented in your manuscripts. • The format of reference citation varies with different journals. Commercially available reference managing softwares, such as EndNote, provide options for choosing the reference citation format for almost all major biomedical journals. • Most journals do not specify the maximum number of references to be allowed in a manuscript. The number of references in a biomedical research article typically range from 20 to 100.
Tables, Illustrations, and Legends to Illustrations	• Tables and illustrations should be prepared separately in good quality. • Color illustrations are usually free of charge for online publication; however, authors must pay costs of reproducing the color illustrations in the print version of the journal. • Illustrations from the same experiment or the same group of experiments may be grouped together in a composite illustration. • Tables and illustrations are uploaded as separate files during online submission. • Commercially available softwares, such as Microsoft Excel and GraphPad Prism, are frequently used for creating graphs and charts from scientific data. • Legends to illustrations should be written on separate sheets immediately after the Reference section. • The legends should provide enough detail for the reader to understand the illustrations without referring back to the Result section. The legends should also describe the statistical significance.
Supplementary Materials	• This is the place where additional data may be included to support your manuscript or to help the reviewers understand your work better. • These data, however, will not be published as part of the manuscript.

3.4.5.3. Submitting the Manuscript to the Journal and Waiting for the Editor's Decision

Finishing writing your manuscript does not necessarily mean that it is ready for submission to the journal. It is advised that you ask colleagues to proofread your manuscript to eliminate typos and grammatical errors. If you are a new investigator, you may also need to ask an experienced researcher to review your manuscript. Once you have finalized your manuscript, it is also important for all the authors to receive a copy of it. Many journals require that the corresponding author state in the cover letter that all authors have read the manuscript and agreed with its submission for publication. The criteria for authorship are discussed in Section 4.3.

Nowadays, almost all biomedical journals require online submission of the manuscripts. Make sure that you strictly adhere to the journal's guidelines for online submission. Following a successful online submission, you should receive an acknowledgment from the journal's editorial office indicating that the office has received your manuscript and specifying the assigned editor or associate editor for handling your manuscript. For high-tier journals, due to the large volume of submissions, the editor usually makes a quick editorial review of the manuscript to see if it reaches the priority for sending out for peer-review. If not, the manuscript will be rejected. Otherwise, the manuscript will be sent out for peer-review usually by 2-3 experts in the field. It usually takes multiple weeks for the reviewers to complete the review and send the comments to the editor. Following examining the reviewers' comments and evaluating the manuscript by him/herself, the editor will make a decision regarding the suitability for the manuscript to be published in the journal. The decision may be one of the following:

- Rejection: This means that the manuscript cannot be considered for publication in the journal. As such, the corresponding author needs to choose another journal for publishing the work.
- Revision: This implies that the manuscript may be considered for publication only after revision to adequately address the reviewers' comments. The revision can be either minor or major. Minor revision usually does not require additional experiments, whereas major revision frequently does.
- Acceptance: The manuscript is acceptable for publication in its present format. This, however, occurs very rarely.

3.4.5.4. Revising the Manuscript and Resubmitting it to the Journal and Waiting for Editor's Decision

Following the initial submission, acceptation of a manuscript without revision for publication in a journal, especially a high-tier one occurs very rarely. Hence, almost all published manuscripts had gone through revision (sometimes multiple rounds of revision) before being accepted for publication. In revising your manuscript, make sure that you respond to the reviewers' comments in a serious and constructive manner. Merely rebutting the reviewers' critiques would diminish the chance for the manuscript to be accepted as the initial reviewers are frequently asked to review the revised manuscript again. On the other hand, if you make the requested revisions, your revised manuscript has a high likelihood of being accepted for publication. However, it is important to also note that if you feel that your manuscript was misinterpreted or misunderstood by the reviewers and the reviewers' comments were not reasonable don't be afraid to argue against the reviewers' criticism.

4. Defining Biomedical Research Ethics

The research community and the community at large rightly expect adherence to exemplary standards of intellectual honesty in the formulation, conduct, and reporting of scientific research. This is particularly true for biomedical research as it is one of the main domains of scientific research that impact greatly on our society. This section is intended to define biomedical research ethics and introduce the basic concepts related to biomedical research ethics.

4.1. Definition of Biomedical Research Ethics

Biomedical research ethics can be broadly defined as application of fundamental ethical principles to the entire process of biomedical research, ranging from planning and conducting research to reporting and reviewing research data. According to this definition, the scope of biomedical research ethics includes the following two categories [36]:

(1) Responsible planning and conducting of biomedical research

- The protection of human subjects in human subjects research (see Chapter 8)
- The welfare of laboratory animals in biomedical research involving the use of animals (see Chapter 7)
- Responsible data management
- Conflicts of interest
- Others

(2) Responsible publication of biomedical research data including publication ethics

- Authorship
- Peer-review
- Others

As the protection of human subjects and the welfare of laboratory animals in biomedical research are covered in additional separate chapters, this section focuses on describing several general terms and concepts, including research integrity and research misconduct as well as publication ethics. These terms and concepts are closely related with extensive overlaps among them.

4.2. Research Integrity and Research Misconduct

4.2.1. Research Integrity

Maintaining a high standard of research integrity is essential to the success of the biomedical research enterprise and benefits the public health and society in general. In the United States, the Office of Research Integrity (ORI) (http://www.ori.hhs.gov) is one of the bodies concerned with research integrity. It was created when the Office of Scientific Integrity (OSI) in the National Institutes of Health (NIH) and the Office of Scientific Integrity Review (OSIR) in the Office of the Assistant Secretary for Health, Department of Health and Human Services (HHS) merged in May 1992. The Office of Research Integrity (ORI) oversees and directs Public Health Service (PHS) research integrity activities on behalf of the Secretary of Health and Human Services with the exception of the regulatory research integrity activities of the Food and Drug Administration (FDA). To maintain research integrity by adhering to federal policies (see below), institutions of higher education as well as many other research institutes/organizations in the United States have established their own offices of research integrity (also see Chapter 6).

Although the term research integrity (also known as scientific integrity) is frequently encountered in the scientific community and public media, there is no unified definition for it. According to the website of ORI (http://www.ori.hhs.gov), research integrity is defined simply as active adherence to the ethical principles and professional standards essential for the responsible practice of research. Here active adherence means adoption of the principles and practices as a personal credo, not simply accepting them as impositions by rule makers, and ethical principles refer to honesty, the golden rule, trustworthiness, and high regard for the scientific record. The golden rule (also known as ethic of reciprocity) here refers to either of the following: (1) one should treat others as one would like others to treat himself/herself, and (2) one should not treat others in ways that he/she would not like to be treated.

The NIH Office of Extramural Research (OER) defines research integrity to include the following three components:

- using honest and verifiable methods in proposing, performing, and evaluating research;
- reporting research results with particular attention to adherence to rules, regulations, guidelines; and
- following commonly accepted professional codes or norms.

Perhaps, the most comprehensive definition for research integrity is the one from the United States National Academy of Sciences. It defines research integrity as follows [37]:

- Integrity characterizes both individual researchers and the institutions in which they work. For individuals, it is an aspect of moral character and experience. For institutions, it is a matter of creating an environment that promotes responsible conduct by embracing standards of excellence, trustworthiness, and lawfulness that inform institutional practices.
- For the individual scientist, integrity embodies a commitment to intellectual honesty and personal responsibility for one's actions and to a range of practices that characterize responsible research conduct. These practices include:

    - intellectual honesty in proposing, performing, and reporting research;
    - accuracy in representing contributions to research proposals and reports;
    - fairness in peer review;
    - collegiality in scientific interactions, including communications and sharing of resources;
    - transparency in conflicts of interest or potential conflicts of interest;
    - protection of human subjects in the conduct of research;
    - humane care of animals in the conduct of research; and
    - adherence to the mutual responsibilities between investigators and their research teams.

Maintaining high standards of research integrity is essential to the success of biomedical research enterprise and benefiting the public health and the society in general. To sustain a culture of scientific integrity, in responding to an initial call by President Barack Obama, the Office of Science and Technology Policy of the White House recently released policies on scientific integrity [38]. This was followed by the release of the scientific integrity policies from the individual departments and agencies across the federal government to guide their diverse missions [39]. These policies along with those adopted at the individual institutions throughout the nation are essential for preventing and minimizing research misconduct in the United States.

4.2.2. Research Misconduct

On December 6, 2000, the Office of Science and Technology Policy in the White House of the United States published the Federal Research Misconduct Policy [40], which required all federal agencies or departments supporting intramural or extramural research to implement it within one year through either policies or regulations. According to the Federal Research Misconduct Policy, research misconduct is defined as fabrication, falsification, or plagiarism (FFP) in proposing, performing, or reviewing research, or in reporting research results.

- Fabrication is making up data or results and recording or reporting them.
- Falsification is manipulating research materials, equipment, or processes, or changing or omitting data or results such that the research is not accurately represented in the research record.
- Plagiarism is the appropriation of another person's ideas, processes, results, or words without giving appropriate credit.
- Research misconduct does not include honest error or differences of opinion.

In the context of this federal policy, research includes all basic, applied, and demonstration research in all fields of science, engineering, and mathematics. This includes, but is not limited to, research in economics, education, linguistics, medicine, psychology, social sciences, statistics, and research involving human subjects or animals.

Institutional research misconduct policies generally follow the pattern recommended by the federal government, but almost always include some additional elements that for one reason or another are assumed to have local importance. This is particularly true for the definition of research misconduct. Institutional definitions must include some version of FFP (fabrication, falsification, or plagiarism), but they sometimes add other practices that also constitute misconduct in the particular local setting. Thus, depending on where a researcher works, any of the following practices could be reported as misconduct in research [36].

- Violation of federal rules
- Abuse of confidentiality
- Authorship and publication violations
- Failure to report misconduct
- Obstruction of investigations and retaliation
- Other practices that seriously deviate from the commonly accepted practices

4.2.3. Research Misconduct Allegations

The Federal Research Misconduct Policy [40] also includes guidelines on what must be present in order to find a researcher guilty of committing research misconduct. According to this policy, a finding of research misconduct requires that:

- there be a significant departure from accepted practices of the relevant research community; and
- the misconduct be committed intentionally, or knowingly, or recklessly; and
- the allegation be proven by a preponderance of evidence.

The National Instituttes of Health (NIH) also recently released a Guide to the Handling of Research Misconduct Allegations [41]. The guide is intended to enable allegations of research misconduct to be processed fairly, confidentially, and promptly. Allegations of research misconduct are taken seriously by the NIH. The process of investigating allegations must be balanced by equal concern for protecting the integrity of the research as well as the careers and reputations of researchers [41].

4.3. Responsible Conduct of Research

With increasing public support for research has come an understandable concern about the way research is conducted. Indeed, public funds support roughly one-third of all research and development in the United States and half of all basic research. Many researchers, therefore, spend a significant portion of their time working for the public. As public servants and also professionals, researchers have clear obligations to conduct their research in a responsible manner.

In general terms, responsible conduct in research is simply good citizenship applied to professional life. Researchers who report their work honestly, accurately, efficiently, and objectively are on the right road when it comes to responsible conduct, and anyone who is dishonest, knowingly reports inaccurate results, wastes funds, or allows personal bias to influence scientific findings is not [36].

There is no one best way to undertake research, no universal method that applies to all scientific investigations. Accepted practices for the responsible conduct of research can and do vary from discipline to discipline and even from laboratory to laboratory. There are, however, some important shared values for the responsible conduct of research that bind all researchers together, such as [36]:

- Honesty: conveying information truthfully and honoring commitments
- Accuracy: reporting findings precisely and taking care to avoid errors
- Efficiency: using resources wisely and avoiding waste
- Objectivity: letting the facts speak for themselves and avoiding improper bias

Adherence to these fundamental values and other widely accepted practices is essential to maintaining research integrity by avoiding research misconduct. In this context, multiple well-recognized materials and programs are available for enhancing education in the responsible conduct of research [15, 36, 37, 42-47]. Listed in **Table 1-12** are three notable publications that address responsible conduct of research.

To help the readers gain a better understanding of the literature on biomedical research ethics, **Box 1-8** lists additional terms on research ethics that are commonly encountered in the literature and public media.

The basic principles of biomedical research process also apply to educational research [56]. It is also for this reason that successful planning and implementation of a biomedical educational research project require the coordinated efforts of a research team that includes educational experts, research scientists, statisticians, and technical assistants among others. Lastly, it is important to distinguish educational research from research education. In this regard, the NIH Research Education (R25) grant mechanism is designed to support the development of creative and innovative research education programs for the development of biomedical, behavioral, and clinical researchers, or for public education and outreach on health-related research to a variety of audiences. Although research education grants are not typical research instruments, they do involve experiments in education and/or dissemination of research knowledge that require an evaluation plan in order to determine their effectiveness [57].

References

[1] Altbach PG, Salmi J. The Road to Academic Excellence: The Making of World-Class Research Universities (Directions in Development) (2011), The World Bank, Washington, D.C.

[2] Explorable.com. Research process (2008). http://explorable.com/research-process.html (accessed November 30, 2012).

[3] Bush V. Science: The Endless Frontier: A Report to the President by Vannevar Bush, Director of the Office of Scientific Research and Development, July 1945 (United States Government Printing Office, Washington: 1945) http://www.nsf.gov/od /lpa/nsf50/vbush1945.htm (accessed November 30, 2012).

[4] Collins FS. NIH basics. *Science.* 2012; 337:503.

[5] Lawrence Berkeley National Laboratories. What is basic research? http://www.lbl.gov/Education/ELSI/Frames/research-basic-defined-f.html (accessed November 30, 2012).

[6] Collins FS. Congressional Justification of the NIH fiscal year (FY) 2013 budget request. http://officeofbudget.od.nih.gov/pdfs/FY13/FY2013_Overview.pdf (accessed November 30, 2012).

[7] Lawrence Berkeley National Laboratories. What is applied research? http://www.lbl.gov/Education/ELSI/research-main.html (accessed November 30, 2012).

[8] The Free Dictionary. Biomedicine. http://medical-dictionary.thefreedictionary.com /biomedicine (accessed November 30, 2012).

[9] Zerhouni EA. US biomedical research: basic, translational, and clinical sciences. *JAMA.* 2005; 294:1352-8.

[10] Perzynski AT. Multidisciplinary approaches to biomedical research. *JAMA.* 2010; 304:2243-4; author reply 2244.

[11] Schaefer GO, Emanuel EJ, Wertheimer A. The obligation to participate in biomedical reserach. *JAMA.* 2009; 302:67-72.

[12] Smith U, Sipido K, Dive C, Nicod L. Alliance for biomedical research in Europe. *EMBO Mol Med.* 2011; 3:505-6.

[13] Moses H 3rd, Martin JB. Biomedical research and health advances. *N Engl J Med.* 2011; 364:567-71.
[14] European Medical Research councils (EMRC). A Stronger Biomedical Research for a Better European Future (2011). http://www.esf.org/uploads/media/emrc_wpII.pdf (accessed November 30, 2012).
[15] World Health Organization. Handbook: Quality Practices in Basic Biomedical Research (2006). http://www.who.int/tdr/publications/documents/quality_practices.pdf (accessed November 30, 2012).
[16] Botstein D. Why we need more basic biology research, not less. *Mol Biol Cell.* 2012; 23:4160-1.
[17] Gura T. US medical researchers brace themselves for budget cuts. *Lancet.* 2012; 380:1371-2.
[18] Kaiser J. Biomedical research policy: NIH funding shifts with disease lobbying, study suggests. *Science.* 2012; 338:181.
[19] NIH. Clinical research. http://grants.nih.gov/grants/policy/hs/glossary.htm (accessed November 30, 2012).
[20] NIH. Executive Summary: NIH Director's Panel on Clinical Research Report 12/97. http://www.oenb.at/de/img/executive_summary--nih_directors_panel_on_clinical_research_report_12_97_tcm14-48582.pdf (accessed November 30, 2012).
[21] Rubio DM, Schoenbaum EE, Lee LS, Schteingart DE, Marantz PR, Anderson KE, Platt LD, Baez A, Esposito K. Defining translational research: implications for training. *Acad Med.* 2010; 85:470-5.
[22] Butler D. Translational research: crossing the valley of death. *Nature.* 2008; 453:840-2.
[23] NIH. Clinical trial, as pertains to the review of applications submitted to the NIH. http://grants1.nih.gov/grants/policy/hs/glossary.htm?print=yes& (accessed November 30, 2012).
[24] NCATS. Institutional Clinical and Translational Science Award (U54). http://grants.nih.gov/grants/guide/rfa-files/rfa-tr-12-006.html (accessed November 30, 2012).
[25] NCI. Transforming translation: harnessing discovery for patient and public health (2007). http://www.cancer.gov/researchandfunding/trwg/finalreport.pdf (accessed November 30, 2012).
[26] Gelijns AC, Gabriel SE. Looking beyond translation: integrating clinical research with medical practice. *N Engl J Med.* 2012; 366:1659-61.
[27] Kaitin KI. Translational research and the evolving landscape for biomedical innovation. *J Investig Med.* 2012; 60:995-8.
[28] Ehrlich E./United for Medical Research. An economic engine: NIH research, employment, and the future of the medical innovation sector (2011). http://www.united formedicalresearch.com/wp-content/uploads/2012/07/UMR_Economic-Engine.pdf (accessed November 30, 2012).
[29] Jekel JF, Katz DL, Elmore JG, Wild DMG. Epidemiology, Biostatistics, and Preventive Medicine, 3rd Edition, 2007, Elsevier, PA.
[30] Gordis L. Epidemiology, 4th Edition, 2009, Elsevier, PA.
[31] NIAID. Design a project. http://www.niaid.nih.gov/researchfunding/grant/strategy /pages/2designproj.aspx (*accessed* March 24, 2013).

[32] Weber RJ, Cobaugh DJ. Developing and executing an effective research plan. *Am J Health Syst Pharm.* 2008; 65:2058-65.

[33] Hanson BP. Designing, conducting and reporting clinical research: a step by step approach. *Injury*. 2006; 37:583-94.

[34] CFR - Code of Federal Regulations Title 21, Part 58: Good Laboratory Practice for Nonclinical Laboratory Studies. http://www.accessdata.fda.gov/scripts/cdrh/cfdocs /cfcfr/cfrsearch.cfm?cfrpart=58 (accessed March 24, 2013).

[35] ICH E6(R1) Good Clinical Practice. http://www.ich.org/products/guidelines /efficacy /efficacy-single/article/good-clinical-practice.html (accessed March 24, 2013).

[36] Steneck NH. ORI Introduction to the Responsible Conduct of Research (2007). http://ori.hhs.gov/documents/rcrintro.pdf (accessed November 30, 2012).

[37] Integrity in Scientific Research: Creating an Environment that Promotes Responsible Conduct (2002), National Academies Press, Washington, D.C. http://www.nap.edu /catalog/10430.html (accessed November 30, 2012).

[38] Office of Science and Technology Policy. Scientific Integrity: Fueling Innovation, Building Public Trust. http://www.whitehouse.gov/sites/default/files/microsites /ostp/scientific-integrity-memo-12172010.pdf (accessed November 30, 2012).

[39] Office of Science and Technology Policy. Scientific Integrity. http://www.whitehouse. gov/administration/eop/ostp/library/scientificintegrity (accessed November 30, 2012).

[40] Office of Science and Technology Policy. Federal Policy on Research Misconduct (2002). http://www.aps.org/policy/statements/upload/federalpolicy.pdf (accessed November 30, 2012).

[41] NIH. A Guide to the Handling of Research Misconduct Allegations (2012). http://sourcebook.od.nih.gov/resethicscases/NIH%20Misconduct2.pdf (accessed November 30, 2012).

[42] NIH. The Guidelines for the Conduct of Research in the Intramural Research Program at NIH (2007). http://sourcebook.od.nih.gov/ethic-conduct/conduct%20research%206-11-07.pdf (accessed November 30, 2012).

[43] On being a Scientist: a Guide to Responsible Conduct in Research (2009), 3rd edition, National Academies Press, Washington, D.C.

[44] Titus S, Bosch X. Tie funding to research integrity. *Nature.* 2010; 466:436-7.

[45] Koocher GP, Keith-Spiegel P. Peers nip misconduct in the bud. *Nature.* 2010; 466:438-40.

[46] Titus SL, Wells JA, Rhoades LJ. Repairing research integrity. *Nature*. 2008; 453:980-2.

[47] Graf C, Wager E, Bowman A, Fiack S, Scott-Lichter D, Robinson A. Best practice guidelines on publication ethics: a publisher's perspective. *Int J Clin Pract Suppl.* 2007; (152):1-26.

[48] NIH. Conflicts of interest. http://ethics.od.nih.gov/topics/coi.htm (accessed November 30, 2012).

[49] Ledford H. NIH finalizes conflict-of-interest rules. published online 23 August 2011 | Nature | doi:10.1038/news.2011.499. http://www.nature.com/news/2011/110823/full /news.2011.499.html (accessed November 30, 2012).

[50] ICMJE. Uniform requirements for manuscripts submitted to biomedical journals: ethical considerations in the conduct and reporting of research: authorship and contributorship. http://www.icmje.org/ethical_1author.html (accessed November 30, 2012).

[51] NIH. Glossary and acronym list. http://grants.nih.gov/grants/glossary.htm#P (accessed November 30, 2012).
[52] Meyers FJ, Pomeroy C. Creating the future biomedical research workforce. *Sci Transl Med.* 2011; 3:102fs5.
[53] Kaye J, Curren L, Anderson N, Edwards K, Fullerton SM, Kanellopoulou N, Lund D, MacArthur DG, Mascalzoni D, Shepherd J, Taylor PL, Terry SF, Winter SF. From patients to partners: participant-centric initiatives in biomedical research. *Nat Rev Genet.* 2012; 13:371-6.
[54] Kaiser J. Biomedical research policy. NIH funding shifts with disease lobbying, study suggests. *Science.* 2012; 338:181.
[55] Postlethwaite TN. Educational Research: Some Basic Concepts and Terminology, 2005, INESCO/International Institute for Educational Planning, Paris.
[56] Horn C, Plazas Snyder B, Coverdale JH, Louie AK, Roberts LW. Educational research questions and study design. *Acad Psychiatry*. 2009; 33:261-7.
[57] NIH. Education Research Grant (R25) mechanism. http://grants.nih.gov/grants /funding/ac_search_results.htm?text_curr=r25&Search.x=0&Search.y=0&Search_Type =Activity (accessed March 24, 2013).

Self-Assessment Questions

[1] Biomedical research is a vast field of science that includes parts of life, physical, and social sciences. It is commonly divided into three intertwined categories:

A. Basic, translational, and clinic
B. Clinical trial, case-control, and behavioral
C. Experimental, clinical, and observational
D. Translational, clinical, and epidemiological

[2] In 1997, the U.S. National Institutes of Health (NIH) Director's Panel on Clinical Research issued a three-part definition of clinical research, which has become widely accepted. According to the NIH definition of clinical research, which of the following is a type of clinical research?

A. Epidemiological studies
B. Studies involving non-human primates
C. Studies involving the use of rats with clinical implications
D. Translational research

[3] Behavioral studies are classified by the U.S. National Institutes of Health as which of the following?

A. Basic biomedical research
B. Clinical research
C. Non-biomedical research
D. Translational research

[4] With regard to biomedical research in the United States, support by the National Institutes of Health (NIH) accounts for more than 75% of the total federal funds devoted to biomedical research. Over 135 Nobel Prizes were won by NIH-supported basic biomedical research scientists. Which of the following is not an appropriate description of NIH?

A. NIH is made up of 27 institutes and centers, each with a specific research agenda, often focusing on particular diseases or body systems.
B. NIH is the directing and coordinating authority on international health within the United Nations' system.
C. NIH is the largest source of funding for medical research in the world, creating hundreds of thousands of high-quality jobs by funding thousands of scientists in universities and research institutions in every state across America and around the globe.
D. NIH, a part of the U.S. Department of Health and Human Services (HHS), is the nation's medical research agency-making important discoveries that improve health and save lives.
E. The current director is Francis S. Collins, MD, PhD

[5] Which of the following belongs to the category of clinical research as defined by the U.S. National Institutes of Health?

A. A study aimed to determine if a human disease can be replicated in monkeys
B. A study in rats that investigates the cause of human hepatocellular carcinoma
C. A study that determines the growth of *H. pylori* (bacteria that cause human peptic ulcer disease) in rabbits
D. A study that seeks to identify the most effective and most efficient health service
E. A study that seeks to identify a new transcriptional activator in a cultured human brain cancer cell line, SH-SY-5Y

[6] The U.S. Food and Drug Administration (FDA) is

A. An agency within the U.S. Department of Health and Human Services with the responsibilities including protecting the public from electronic product radiation
B. Made up of 16 centers, each with a specific research agenda, often focusing on particular drugs or food products
C. Part of the U.S. Center for Disease Control and Prevention (CDC)
D. The largest source of funding for medical research in the world, creating hundreds of thousands of high-quality jobs by funding thousands of scientists in universities and research institutions in every state across America and around the globe

[7] Clinical trials used in drug development are described in phases. According to the U.S. Food and Drug Administration, which of the following defines phase 0 clinical trial?

A. A trial that is done after a new drug has been marketed

B. A trial that is done to study the biomedical or behavioral intervention in a larger group of people (several hundred) to determine efficacy and to further evaluate its safety
C. A trial that is done to study the efficacy of the biomedical or behavioral intervention in large groups of human subjects (from several hundred to several thousand) by comparing the intervention to other standards or experimental interventions as well as to monitor adverse effects, and to collect information that will allow the intervention to be used safely
D. Exploratory study involving very limited human exposure to the drug, with no therapeutic or diagnostic goals

[8] Professor T.M. and a DO/PhD candidate propose a clinical study involving 22 healthy volunteers to determine if "BrocoProtector", a new chemical recently isolated by Professor T.M. from broccoli, is safe for oral intake at a dose of 0.5 grams daily for 7 days. This clinical study falls into which of the following phases?

A. Phase I
B. Phase II
C. Phase III
D. Phase IV

[9] Professor G.T. lectures on translational research design to a small class of first year medical students. A student asks if investigation of factors that influence the health of populations, which ultimately results in improved health of the public is called translational research. Professor G.T. replies with Yes and further states that this is one of the 4 phases of translation defined by the recently established U.S. National Center for Advancing Translational Sciences (NCATS). Which phase does Professor G.T. refer to?

A. T1 translation
B. T2 translation
C. T3 translation
D. T4 translation

[10] A clinical scientist teaches a clinical research design course to a family medicine residency program. She assigns a recent article published in New England Journal of Medicine as a required reading material. Upon reading the article, one resident doctor realizes that this article reports a double-blind randomized clinical trial on a new drug that targets the renin-angiotensin-aldosterone axis in diabetic patients with high blood pressure. Here "double-blind" means:

A. A type of masking in which neither of the two parties (e.g., investigators and patients) involved in the trial knows which participants have been assigned which intervention.
B. This trial deals with drug effects on vision.
C. Two of the investigators do not know anything about the trial.
D. Two of the participants (diabetic patients with hypertension) are blind.

[11] Most scientific research seeks to identify causal relationships. Indeed, investigation of the causal relationships is one of the most important and challenging tasks of both basic and clinical biomedical research. Risk factor is one of the three fundamental types of causal relationships. Which of the following best describes the term risk factor?

A. Cigarette smoking is a risk factor for lung cancer
B. Drinking coffee is a risk factor for lung cancer
C. Homozygous hexosaminidase A deficiency is a risk factor for Tay-Sachs disease
D. *M. Tuberculosis* is a risk factor for tuberculosis

[12] Bellemain-Appaix et al. recently published an article in JAMA (2012; 308:2507-2516), entitled: "Association of clopidogrel pretreatment with mortality, cardiovascular events, and major bleeding among patients undergoing percutaneous coronary intervention: a systematic review and meta-analysis." Meta-analysis is:

A. A statistical process that combines the findings from many individual studies and it provides much stronger evidence on causal relationship than individual trials.
B. A type of observational study in which the investigators select a group of exposed individuals and a group of non-exposed individuals and then follow up both groups to compare the incidence of disease.
C. A type of observational study that compares two groups of people: those with the disease or condition under study (cases) and a very similar group of people who do not have the disease or condition (controls)
D. A cross-sectional study that measures the prevalence of health outcomes or determinants of health, or both, in a population at a point in time or over a short period.

[13] The Transient Ischemic Attack and Stroke Patients and Unelucidated Risk Factor Evaluation (FUTURE) study is a prospective cohort study of prognosis after transient ischemic attack (TIA), ischemic stroke, or hemorrhagic stroke in adults aged 18 through 50 years admitted to Radboud University Nijmegen Medical Centre, the Netherlands, between January 1, 1980, and November 1, 2010. A recent follow-up of the FUTURE study was published in JAMA (2013; 309:1136-1144). Which of the following best defines a cohort study?

A. A double-blind study that measures the prevalence of health outcomes or determinants of health, or both, in a population at a point in time or over a short period
B. A single-blind randomized controlled trial that determines the consequence after a major cerebral vascular event
C. A type of observational study in which the investigators select a group of exposed individuals and a group of non-exposed individuals and then follow up both groups to compare the incidence of disease.
D. A type of observational study that compares two groups of people: those with the disease or condition under study (cases) and a very similar group of people who do not have the disease or condition (controls)

[14] The U.S. Food and Drug Administration (FDA) Code of Federal Regulations 21CFR58: Good Laboratory Practice (GLP) for Preclinical Studies prescribes good laboratory practices for conducting nonclinical laboratory studies that support or are intended to support applications for research or marketing permits for products regulated by the FDA. Which of the following best describes the principles of GLP?

A. CLP is an international ethical and scientific quality standard for designing, conducting, recording, and reporting trials that involve the participation of human subjects.
B. GLP helps assure regulatory authorities that the data submitted are a true reflection of the results obtained during the study and can therefore be relied upon when making risk/safety assessments.
C. GLP is consistent with the principles that have their origin in the Declaration of Helsinki, and that the clinical trial data are credible.
D. GLP is not recommended for laboratory studies that are not regulated by federal agencies.

[15] Which of the following types of clinical design would produce the strongest evidence for a possible causal relationship?

A. Case report
B. Case-control
C. Cross-sectional study
D. Prospective cohort

[16] Journal impact factor is a measure of the frequency with which the "average article" in a journal has been cited in a given period of time. Which of the following best describes the characteristics of journal impact factor?

A. A perfect way of assessing the value or quality of a biomedical journal
B. Doesn't change from year to year for a particular journal
C. General journals tend to have higher impact factors than subspecialty journals. However, subspecialty journals are more likely to have a more focused audience.
D. Journals that publish more review articles tend to have lower impact factors than those that only publish research articles.
E. Thomson Reuters depends on the impact factor alone in assessing the usefulness of a journal.

[17] Biomedical research ethics refers to application of fundamental ethical principles to the entire process of biomedical research, ranging from planning and conducting research to reporting and reviewing research data. In this regard, The U.S. National Institutes of Health (NIH) Office of Extramural Research (OER) defines research integrity to include three components. Which of the following is not one of the three components?

A. Following commonly accepted professional codes or norms
B. Good Laboratory Practice

C. Reporting research results with particular attention to adherence to rules, regulations, and guidelines
D. Using honest and verifiable methods in proposing, performing, and evaluating research

[18] According to the U.S. Federal Research Misconduct Policy, research misconduct is defined as fabrication, falsification, or plagiarism (FFP) in proposing, performing, or reviewing research, or in reporting research results. Which of the following is considered plagiarism?

A. Differences of opinion
B. Making up data or results and recording or reporting them
C. Manipulating research materials, equipment, or processes, or changing or omitting data or results such that the research is not accurately represented in the research record
D. The appropriation of another person's ideas, processes, results, or words without giving appropriate credit

[19] GS, a graduate student in Professor D.J.'s lab has just finished writing a manuscript based on his thesis research. The authorship includes himself as the first author, a technician working in the lab as the second author, and Professor D.J. as the last and also corresponding author. Professor D.J. gives a copy of the manuscript to a postdoctoral fellow in his lab for review. Upon seeing his name is not on the paper, the fellow gets upset and insists that his name be on the paper because he has shown the graduate student how to run an assay that has been subsequently used by the graduate student to generate several pieces of data for the manuscript. Professor D.J. kindly provides the postdoctoral fellow a copy of the guidance for authorship from the International Committee of Medical Journal Editors (ICMJE) and asks him to read through the authorship requirement section. Two days later, the fellow apologizes to Professor D.J. for his inappropriate demand. According to the ICMJE, an author is generally considered to be someone who has made substantive intellectual contributions to a published study. Authorship credit should be based on three conditions, and authors should meet all the three conditions. Which of the following is not one of the three conditions?

A. Drafting the article or revising it critically for important intellectual content
B. Final approval of the version to be published
C. Having a doctorate degree
D. Substantial contributions to conception and design, acquisition of data, or analysis and interpretation of data

[20] Evidence-based practice (EBP) is a commonly encountered term in today's practice of medicine and patient care. EBP is the integration of clinical expertise, patient values, and the best research evidence into the decision making process for patient care. EBP requires new skills of the clinician, including efficient literature searching, and the

application of formal rules of evidence in evaluating the clinical literature. Which of the following is an example of evidence-based practice?

A. A clinician recommends the use of a non-statin drug to treat his patient's hypercholesterolemia because one of his family members responds well to the non-statin drug.
B. A clinician recommends the use of a newly approved drug for treating his patient's hypertension despite his patient responds well to the current therapy. The physician believes that new drugs should always work better than the old ones.
C. A clinician wanted to know about the risk of stomach bleeding in people taking aspirin. He did a meta-analysis of data from 24 clinical trials with nearly 66,000 participants and found that the risk of stomach bleeding was 2.47 percent with aspirin compared to 1.42 percent with placebo (inactive substance). His analysis is an example of evidence-based practice.
D. A health care provider recommends acetaminophen to treat arthritis pain in a patient who has recently had stomach bleeding. The health care provider makes this recommendation because research shows that acetaminophen is associated with less risk for stomach bleeds than other common pain relievers. The health care provider's recommendation is an example of evidenced-based practice.

Chapter 2

The Essential Nature of Biomedical Research

Abstract

Biomedical research in the United States is an approximately $140 billion enterprise, and the benefits of biomedical research are enormous. Chapter 1 defines biomedical research and its scope, biomedical research process, and research ethics. This chapter discusses the essentiality of biomedical research from two broad aspects: (i) benefits to public health and social and economic development, and (ii) impact on the prestige and value of institutions of higher education. Biomedical research is essential to institutional prestige and value. The extent to which institutions of higher education, including medical schools are engaged in biomedical research and development activities has a key role in determining the status and quality of these institutions and the contribution which they make to public health, and economic and social development. The essential nature of biomedical research has been also a driving force for establishing biomedical research programs at traditionally non-research-intensive universities, where teaching is the primary focus.

Chapter at a Glance

1. Overview

The biomedical research community has long considered research to be vital to the health and wealth of societies, supporting the view attributed to May Lasker, American philanthropist and ardent campaigner for biomedical research: "if you think research is expensive, try disease" [1]. Biomedical research in the United States is an approximately $140 billion enterprise, with about 56% supported by industry, 30% by federal government (predominantly by the National Institutes of Health which accounts for more than 75% of the total federal funds), and 14% by others (institutional funds, state and local government, charities, foundations, or individual donors) [2, 3]. The nation's sustained investment in biomedical research has made a tremendous impact on the health of human beings of the entire world as well as on the social and economic development of the United States and the rest of the world. This chapter describes the wider benefits of biomedical research, providing a rationale for continued efforts and resources devoted to biomedical research at levels of both societies and individual institutions of higher education. This chapter also discusses the essential nature of biomedical research to the prestige and value of institutions of higher education, which may serve as a rationale for developing biomedical research in traditionally non-research-intensive universities.

2. Role of Biomedical Research in Public Health, National Security, and Social and Economic Development

The benefits of biomedical research have been widely documented. This section examines the impact of biomedical research on public health, national security, and social and economic development with a focus on the impact of biomedical research funded by the United States National Institutes of Health, the world's largest biomedical research agency.

2.1. Impact on Public Health

2.1.1. Overview of the Public Health Benefits of Biomedical Research

Over the years, biomedical research, funded in large part by the United States National Institutes of Health (NIH), has contributed enormously to an increase in health and life expectancy in the U.S. as well as worldwide. As described in the 2008/2009 NIH biennial report to the U.S. Congress, life expectancy increased by 7.4 years from 1961 to 2004. Infant mortality decreased from 26 deaths per 1,000 live births in 1960 to 6.9 in 2005 [4].

In the Congressional Justification of the NIH fiscal year 2013 budget request, NIH director Francis Collin further stated that NIH research has an impressive track record of producing tangible improvements in the diagnosis, treatment, and prevention of disease [5]. Thanks in large part to NIH investments, after three decades of scientific progress in HIV/AIDS prevention and treatment, we can begin to imagine an AIDS-free generation. We have seen, for example, dramatic reductions in death rates for coronary disease and for stroke (declining 60-70 percent in the last half century), tripling of the number of Americans living into their 90s over the past twenty years, and a substantial reduction in disability among all seniors. On the other end of the lifecycle, we have seen a 40 percent decline in infant mortality over 20 years and better treatments for premature and low-weight births [5].

Similarly, because of biomedical research infant mortality in the Europe has, for example, decreased from 28.6 deaths per 1,000 lives in 1965 to 4.7 in 2006. Death rates in Europe had fallen for all main causes of death except one group (i.e., malignant neoplasms) for both men and women between 2000 and 2008 [6].

2.1.2. Fundamental Role of Basic Biomedical Research in Achieving the Public Health Benefits

Biomedical research has expanded and continues to broaden our understanding of the pathophysiology underlying many diseases (often at the molecular level), contributing, along with other factors such as changes in behavior, to numerous advances in treatments and improved health care. The change in the prognosis for HIV patients is one example of these benefits. In the 1990s, the discovery and development of antiretroviral drugs transformed HIV infection for many individuals from a death sentence into a chronic disease [5].

The development of effective drugs has revolutionized the treatment of heart attacks and high blood pressure, and enabled many people with schizophrenia to emerge from mental hospitals to live at home. In the treatment and prevention of cardiovascular diseases, a third of the gains can be attributed to high-tech invasive treatments such as coronary bypass surgery and percutaneous coronary intervention, another third to medications that treat conditions such as hypertension and dyslipidemias, and a final third to behavioral changes achieved through clinical trials revealing evidence for education on smoking, diet, and exercise.

Many people appreciate the advance in medical achievements but remain unaware of the fundamental role of basic biomedical research discoveries leading to the development of new medicines and cures for human diseases. The dramatic reduction in deaths from ischemic coronary disease over the past 40 years, as repeatedly mentioned in this section, is but one example of success resulting largely from NIH-funded research that [7]:

- elucidated the basic biology of cholesterol metabolism leading to the discovery of lipid-lowering medications;
- fueled the development of coronary stents and the substances with which they are coated; and
- encouraged the analysis of interventions that improve clinical practice.

In addition to the improvement of public health, biomedical research has also recently received considerable attention with regard to its impact on national security, especially issues involving chemical and biologic terrorism. The next section explores the significance of biomedical research in the context of national security.

2.2. Impact on National Security [8-12]

Biomedical research is not only vital to the improvement of human health, but is also an integral component of the national security of the United States as well as many other countries throughout the whole world. The terrorist attacks on the World Trade Center and the Pentagon on September 11, 2001, and the deliberate release of anthrax spores that occurred soon thereafter starkly revealed the United States' vulnerability to the threat and reality of multiple categories of terrorism, including biologic, chemical, nuclear, and radiological assaults [8].

As a result of the events of late 2001, the focus on national security in the United States has intensified greatly, and extensive efforts to prepare for and prevent future attacks have been undertaken, particularly by the newly established Department of Homeland Security. In the arena of biologic terrorism against the civilian population, the United States Department of Health and Human Services has taken a lead role. In this context, extensive efforts have focused on surveillance and activities to promote public health preparedness, led by the Centers for Disease Control and Prevention (CDC) (**Box 2-1**), and biomedical research, led by the National Institutes of Health (NIH), particularly the National Institute of Allergy and Infectious Diseases (NIAID) [8]. NIAID is an institute under NIH (see Box 1-4 of Chapter 1 for more information on NIH).

Box 2-1. The United States Center for Disease Control and Prevention (CDC)
(www.cdc.gov)

CDC at a glance:

- CDC was founded in July 1946, and is one of the major operating components of the U.S. Department of Health and Human Services.
- Today, CDC is the nation's premier health promotion, prevention, and preparedness agency and a global leader in public health.
- It remains at the forefront of public health efforts to prevent and control infectious and chronic diseases, injuries, workplace hazards, disabilities, and environmental health threats.
- CDC is globally recognized for conducting research and investigations and for its action-oriented approach. CDC applies research and findings in biomedicine to improve people's daily lives and responds to health emergencies- something that distinguishes CDC from its peer agencies.
- CDC works with states and other partners to provide a system of health surveillance to monitor and prevent disease outbreaks (including bioterrorism), implement disease prevention strategies, and maintain national health statistics. CDC also guards against international disease transmission, with personnel stationed in more than 25 foreign countries.

Mission statement of CDC:

- Collaborating to create the expertise, information, and tools that people and communities need to protect their health – through health promotion, prevention of disease, injury and disability, and preparedness for new health threats.

Strategic areas of CDC:
CDC is now focusing on becoming a more efficient and impactful agency by focusing on five strategic areas:

- Supporting state and local health departments
- Improving global health
- Implementing measures to decrease leading causes of death
- Strengthening surveillance and epidemiology
- Reforming health policies

Director: Thomas R. Frieden, MD, MPH (as of June 2009)

The specific mandate of the NIAID in the post-September 11 national security effort is to support research that will ultimately lead to the development of medical countermeasures in the form of therapies, vaccines, and diagnostic tools to protect the nation from deliberate attacks with biologic agents. In fact, NIAID is the lead component of NIH for research on biodefense against terrorist attacks with pathogenic microbes or toxins and naturally occurring emerging and re-emerging infectious diseases, including seasonal and pandemic influenza [8].

When discussing the involvement of biomedical research in national security, one key issue cannot be neglected, which is the dual use research concern (DURC). As stated by the NIAID director Anthony S. Fauci, MD in his 2012 Testimony before the Committee on Homeland Security and Governmental Affairs, United State Senate, a smaller portion of biological research has a greater potential for yielding knowledge that could be used for harm. This subset of dual use research is known as "dual use research of concern," or DURC. DURC is research that, based on current understanding, can be reasonably anticipated to provide knowledge, information, products, or technologies that could be directly misapplied to pose a significant threat with broad potential consequences to public health and safety, agricultural crops and other plants, animals, the environment, materiel, or national security [9]. Categories of research that should be closely scrutinized for DURC potential include experiments that, for a specific group of agents and toxins, cause the following effects [9]:

- Enhancement of the harmful consequences of an agent or toxin
- Disruption of the effectiveness of an immunization against an agent or toxin
- Conferring resistance to clinically useful prophylactic or therapeutic interventions against an agent or toxin
- Increase of the stability or transmissibility of an agent or toxin
- Alteration of the host range of an agent or toxin
- Enhancement of the susceptibility of a host population to an agent or toxin
- Generation or reconstitution of an eradicated or extinct agent or toxin

However, on the other hand, as stated in the testimony by Dr. Fauci, dual use research ultimately could yield new information critical to the development of technologies needed to improve public health, such as vaccines, diagnostics, and therapeutics. Because of NIAID's lead federal role in supporting and conducting biodefense and emerging infectious diseases research, it can be expected that NIAID has funded and will fund some measure of DURC within its research portfolio. If a particular research experiment is identified as DURC that designation does not necessarily mean that such research should be prohibited or avoided or not widely published. It is imperative to carefully balance the benefit of the science to public health, the biosafety and biosecurity conditions under which the research is conducted, and the potential risk that the knowledge gained from such research may fall into the hands of individuals with ill intent [9, 13].

2.3. Impact on Economic Development

There is now overwhelming evidence that investment in biomedical research yields economic returns, both through improved health gains (a healthy workforce) and through commercial exploitation of research outputs (**Figure 2-1**). In 2008, a United Kingdom (UK) study demonstrated that the health and gross domestic product (GDP) gains derived from the country's public and charitable investments in biomedical research are equivalent to an annual rate of return of about 39% for cardiovascular diseases and 37% for mental health research [6]. This combined an annual rate of return of 30% in GDP gains and of 9% and 7% in health gains from new preventive and therapeutic interventions. The study crucially also showed that the time lag between research funding and health return is approximately 17 years. A similar study performed in the United States found that every U.S. $1 spent by the National Institutes of Health (NIH) typically generates U.S. $2.21 in additional economic output within 12 months [14]. Listed in **Table 2-1** below are examples of the economic benefits of NIH-funded biomedical research.

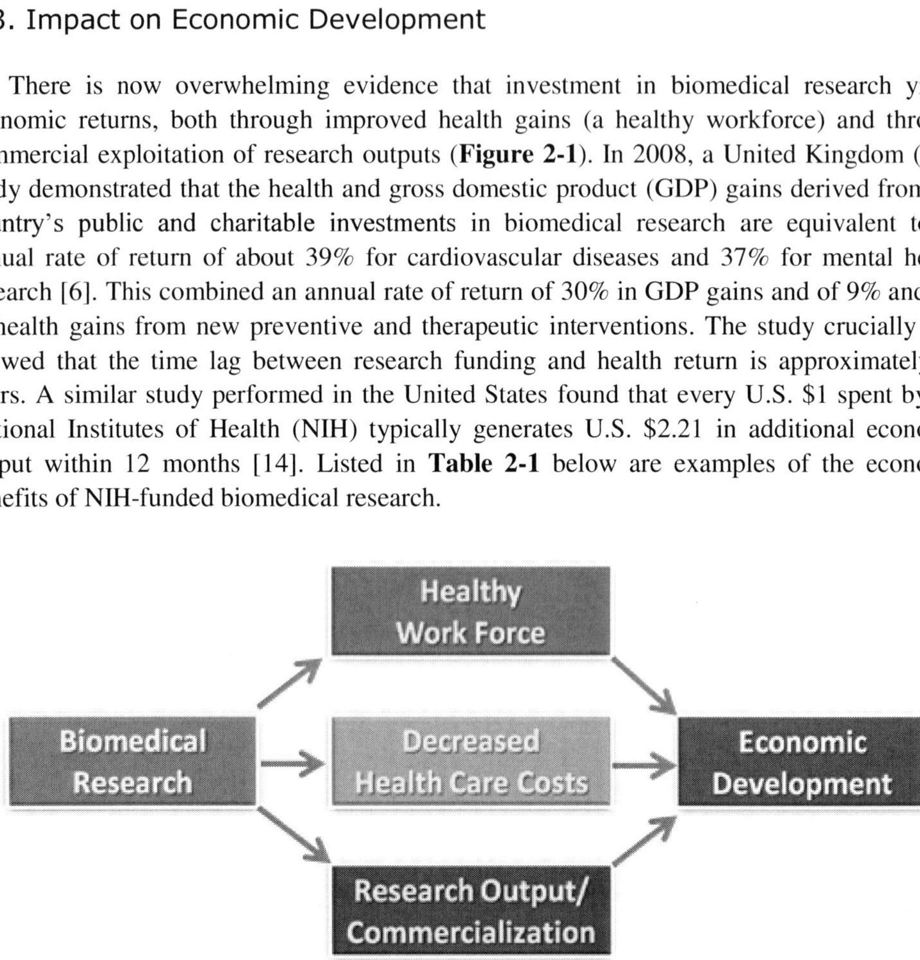

Figure 2-1. Impact of biomedical research on economic development. Biomedical research (predominantly supported by the United States NIH) has resulted in remarkable U.S. gains in health and longevity and healthy work force, often with surprisingly modest investments and often accompanied by significant cost savings [4]. A recent analysis of the trajectory of U.S. population health shows substantial correlation of NIH funding with improved life expectancy, reduced disability rates, and economic benefits. For example, deaths from coronary heart disease have declined by 63 percent in the last 30 years, thanks to a host of new insights about prevention and treatment. These dramatic advances have come about with an investment of just $3.70 per American per year in NIH research support [14a]. Another example of savings we have seen over time is the development of a vaccine against *Haemophilus influenzae* type b (Hib), which has resulted in a 99 percent decline in the incidence of this leading cause of bacterial meningitis in children under age 5. This has achieved an estimated medical cost savings of $950 million per year, as well as another $1.14 billion per year in avoidance of lost earnings due to disability of the patient and uncompensated caregivers [4].

Table 2-1. Examples of the economic benefits of NIH-funded biomedical research

SOURCE OF STUDIES	ECONOMIC BENEFITS
United for Medical Research [15]	NIH is a catalyst and powerhouse for creating and maintaining economic growth across the United States. As documented by the United for Medical Research's An Economic Engine, in 2010, NIH funding: • Led to the creation of 487,900 quality jobs • Produced $68.035 billion in new economic activity • Allowed 16 states to experience job growth of 10,000 jobs or more
Economic Impact of the Human Genome Project [16]	A published report, the Economic Impact of the Human Genome Project, prepared by the Battelle Technology Partnership Practice, estimated: • A stunning $796 billion in economic benefits to the U.S. from the initial federal investment of $3.8 billion in the Human Genome Project. That's a 141 to 1 return on investment. • In 2010 alone, human genome sequencing activities generated $67 billion in U.S. economic output, $20 billion in personal income for Americans, and 310,000 U.S. jobs.

Biomedical research is the core of research at medical schools, and the economic impact of the biomedical research conducted in the United States medical schools is extensive. To add to public understanding of the economic benefits of biomedical research form United States medical colleges, the Association of American Medical Colleges (AAMC) commissioned Tripp Umbach, a highly respected national economic consulting firm, to examine the economic impact of federal- and state-funded medical research conducted at AAMC-member medical schools and teaching hospitals. The Tripp Umbach report shows that federal- and state-funded research received by medical schools and teaching hospitals in 2009 added close to $45 billion to the U.S. economy. To put this in perspective, the National Institute of Health (NIH), the largest federal funding agency of medical research, invested approximately $28.5 billion in fiscal year 2009 for extramural research conducted across the nation, including American Recovery and Reinvestment Act (ARRA) funding. Of those funds, about 55 percent (or about $15.6 billion) went to medical schools and teaching hospitals, a particularly productive research environment where physicians and scientists deliver care to patients, help train the next generation of physicians and researchers, and conduct vital medical research. While NIH funding is critical, the data show that for every institutional dollar invested in research at medical schools and teaching hospitals, $2.60 of economic activity is generated [17]. In addition to the economic benefits of biomedical research in medical schools, the impact of such research on institutional prestige and value is also enormous. This is not only true for medical schools, but also for any other institutions engaged in biomedical research.

3. Impact of Biomedical Research on Institutional Prestige and Value

This section discusses the essential nature of biomedical research to an institution's prestige and value. It begins with an introduction to the methodologies of the three leading international ranking systems of world-class universities, highlighting the core role of research in the ranking of world-class institutions of higher education. This section focuses on the impact of biomedical research as it relates to the prestige and value of medical schools.

3.1. Role of Research in the Ranking of Institutions of Higher Education

Research is generally regarded as a core element of the mission of institutions of higher education. The extent to which universities are engaged in research and development activities has a key role in determining the status and quality of these institutions and the contribution which they make to economic and social development. Research and funding have been widely used as a key criterion in the ranking of world-class institutions of higher education (**Table 2-2**) [18]. As shown in **Table 2-2**, research and publications are weighted heavily for the ranking of world-class universities. World-class universities are exclusively major research institutions.

Table 2-2. Summary of methodologies of the three leading international ranking systems

Ranking System	Indicators/Criteria
Times Higher Education (THE) (http://www.timeshighereducation.co.uk/world-university-rankings) THE ranks 400 world universities. An overall score is calculated for each university employing 13 carefully calibrated performance indicators which are classified into 5 categories. Note: Universities are excluded from the Times Higher Education World University Rankings if they do not teach undergraduates; if they teach only a single narrow subject; or if their research output amounted to fewer than 1,000 articles between 2006 and 2010 (200 a year).	Teaching: the learning environment (worth 30% of the overall ranking score) Research: volume, income and reputation (worth 30%) Citations: research influence (worth 30%) Industry income: innovation (worth 2.5%) International outlook: staff, students and research (worth 7.5%)
The Academic Ranking of World Universities (ARWU) (http://www.arwu.org) ARWU uses six objective indicators to rank 500 world universities. For each indicator, the highest scoring institution is assigned a score of 100, and other institutions are calculated as a percentage of the top score. The distribution of data for each indicator is examined for any significant distorting effect; standard statistical techniques are used to adjust the indicator if necessary. Scores for each indicator are weighted as shown below to arrive at a final overall score for an institution. The highest scoring institution is assigned a score of 100, and other institutions are calculated as a percentage of the top score. An institution's rank reflects the number of institutions that sit above it.	Alumni of an institution winning Nobel Prizes and Fields Medals (10%) Staff of an institution winning Nobel Prizes and Fields Medals (20%) Highly cited researchers in 21 broad subject categories (20%) Papers published in Nature and Science (20%) Papers indexed in Science Citation Index-expanded and Social Science Citation Index (20%) Per capita academic performance of an institution (10%)

RANKING SYSTEM	INDICATORS/CRITERIA
The Higher Education Evaluation and Accreditation Council of Taiwan (HEEACT) (http://www.heeact.edu.tw) HEEACT ranks 500 universities. An overall score is calculated for each university for each of eight indicators; for each indicator, the university with the highest number receives the maximum points; the other universities' numbers are subdivided and converted decimally into their respective scores.	Number of articles in the past 11 years (10%) Number of articles in the current year (10%) Number of citations in the past 11 years (10%) Number of citations in the past 2 years (10%) Average number of citations in the past 11 years (10%) H-index in the past 2 years (20%) Number of highly cited papers (15%) Number of articles in the current year in highly cited journals (15%)

3.2. Biomedical Research at Medical Schools

3.2.1. Introduction to Medical Schools in the United States

Medical schools in the United States include allopathic medical schools which award the doctor of medicine (MD) degree and the osteopathic medical schools which award the doctor of osteopathic medicine (DO) degree. MD-granting medical schools are accredited by the Liaison Committee on Medical Education (LCME), while DO-granting medical schools are accredited by the Commission on Osteopathic College Accreditation (COCA). **Table 2-3** summarizes the major similarities and differences between the MD and DO professions [19]. The core competencies of allopathic and osteopathic professions are listed in **Table 2-4**. There are presently 141 accredited MD-granting institutions and 29 accredited DO-granting institutions operating in 34 locations in the United States. Notably, the number of new DO schools has been growing rapidly in recent years. Due to its focus on primary care, DO profession plays an increasingly important role in health care and has received unprecedented attention and respect in the United States and worldwide in recent years.

Table 2-3. Similarities and differences between doctor of medicine (MD) and doctor of osteopathic medicine (DO) [19]

SIMILARITY	DIFFERENCE
Students entering both DO and MD medical colleges typically have already completed four-year bachelor's degrees with an emphasis on scientific courses.	Osteopathic medical schools emphasize training students to become primary care physicians. Osteopathic physicians focus on preventive health care.
Both DOs and MDs complete four years of basic medical education. After medical school, both DOs and MDs obtain graduate medical education through internships, residencies and fellowships. This training lasts three to eight years and prepares DOs and MDs to practice a specialty.	DOs practice a "whole person" approach to medicine. Instead of just treating specific symptoms or illnesses, they regard your body as an integrated whole. This has become less distinct for DOs as some MDs are also using a whole person approach (see integrative medicine in Box 10-3 of Chapter 10).

Table 2-3. (Continued)

SIMILARITY	DIFFERENCE
Both DOs and MDs can choose to practice in any specialty of medicine-such as pediatrics, family medicine, psychiatry, surgery, or ophthalmology. DOs and MDs must pass comparable examinations to obtain state licenses (USMLE for MD students; COMLEX for DO students; DO students may also take USMLE). DOs and MDs both practice in accredited and licensed health care facilities in all 50 states of the United States. Together, DOs and MDs enhance the state of health care available in the United States.	DOs receive extra training in the musculoskeletal system and osteopathic manipulative treatment (OMT) is incorporated into the training and practice of osteopathic physicians. This is unique for DOs. Many clinical studies demonstrated the beneficial effects of OMT in chronic disorders, especially chronic pain. However, the clinical effectiveness and efficacy of OMT in various disease conditions remain to be further investigated by large-scale randomized controlled trials. In addition, the biological mechanisms of OMT warrant investigation.

Table 2-4. Core competencies for MDs and DOs

ACGME SIX CORE COMPETENCIES (MD) (HTTP://WWW.ACGME.ORG)	AOA/OGME SEVEN CORE COMPETENCIES (DO) (HTTP://WWW.OSTEOPATHIC.ORG)
Beginning in July 2002, the Accreditation Council for Graduate Medical Education (ACGME) instructed all residency programs to require their residents to demonstrate competency in 6 core areas: patient care, interpersonal and communication skills, medical knowledge, professionalism, practice-based learning, and systems-based practice. The goal was to have objective markers of performance that would serve as a gauge to determine a program's accreditation. There are 6 core competencies for MDs, as listed below. **Patient Care:** Identify, respect, and care about patients' differences, values, preferences, and expressed needs; listen to, clearly inform, communicate with and educate patients; share decision making and management; and continuously advocate disease prevention, wellness, and promotion of healthy lifestyles, including a focus on population health. **Medical Knowledge:** Established and evolving biomedical, clinical, and cognate (e.g., epidemiological and social behavioral) sciences and the application of knowledge to patient care.	AOA core competencies were developed by a task force. The task force defined a framework of 7 competency areas for Osteopathic Graduate Medical Education (OGME). These were adopted for the osteopathic profession by AOA board action in July 2003 [21]. Listed below are the 7 core competencies for DOs. **Osteopathic Philosophy and Osteopathic Manipulative Medicine:** Residents are expected to demonstrate and apply knowledge of accepted standards in osteopathic manipulative treatment (OMT) appropriate to their specialty. The educational goal is to train a skilled and competent osteopathic practitioner who remains dedicated to life -long learning and to practice habits in osteopathic philosophy and manipulative medicine. **Medical Knowledge:** Residents are expected to demonstrate and apply knowledge of accepted standards of clinical medicine in their respective specialty area, remain current with new developments in medicine, and participate in life-long learning activities, including research.

ACGME Six Core Competencies (MD)	AOA/OGME Seven Core Competencies (DO)
Practice-Based Learning and Improvement: Involves investigation and evaluation of one's own patient care, appraisal and assimilation of scientific evidence, and improvements in patient care. Additional documentation is required to be awarded AMA PRA Category 1 Credit™ for this ACGME core competency. **Interpersonal and Communication Skills:** That result in effective information exchange and teaming with patients, their families and other health professionals. **Professionalism:** Commitment to carrying out professional responsibilities, adherence to ethical principles and sensitivity to a diverse patient population. **Systems-Based Practice:** Actions that demonstrate an awareness of and responsiveness to the larger context and system of health care and the ability to effectively call on system resources to provide care that is of optimal value. Note: The 7th competency? The medical profession must become more cost-conscious and decrease unnecessary care that does not benefit patients but represents a substantial percentage of health care costs. In this context, there has been discussion on including cost-conscious care as the 7th ACGME core competency for MDs [20].	**Patient Care:** Residents must demonstrate the ability to effectively treat patients, provide medical care that incorporates the osteopathic philosophy, patient empathy, awareness of behavioral issues, the incorporation of preventive medicine, and health promotion. **Interpersonal and Communication Skills:** Residents are expected to demonstrate interpersonal and communication skills that enable them to establish and maintain professional relationships with patients, families, and other members of health care teams. **Professionalism:** Residents are expected to uphold the Osteopathic Oath in the conduct of their professional activities that promote advocacy of patient welfare, adherence to ethical principles, collaboration with health professionals, life-long learning, and sensitivity to a diverse patient population. Residents should be cognizant of their own physical and mental health in order to effectively care for patients. **Practice-Based Learning and Improvement:** Residents must demonstrate the ability to critically evaluate their methods of clinical practice, integrate evidence-based medicine into patient care, show an understanding of research methods, and improve patient care practices. **Systems-Based Practice:** Residents are expected to demonstrate an understanding of health care delivery systems, provide effective and qualitative patient care within the system, and practice cost-effective medicine. Note: The 7 core competencies for DOs cover all the core competencies for MDs in addition to the osteopathic philosophy and manipulative medicine.

3.2.2. Research Activities at Medical Schools

Medical institutions increase in prestige and value to society when they participate in research that results in new knowledge and discoveries that help prevent and manage diseases and promote health, and that contribute to the advances in both basic and clinical sciences (**Figure 2-2**). Indeed, all of the top-tier medical schools in the United States are also major research institutions. Research publications and grant funding are among the key criteria used for ranking medical schools by well-recognized ranking systems, including the U.S. News & World Report ranking of medical schools (**Box 2-2**) [22]. As discussed above, research and publications are also key indicators used for the ranking of world-class institutions of higher education in general. Table 10-1 of Chapter 10 also outlines the major benefits of developing undergraduate biomedical research at institutions of higher education, including both medical schools and non-medical schools that also conduct research in life science. In this context, biomedical research is among the top priorities of research portfolios in the departments of biology of the non-medical schools.

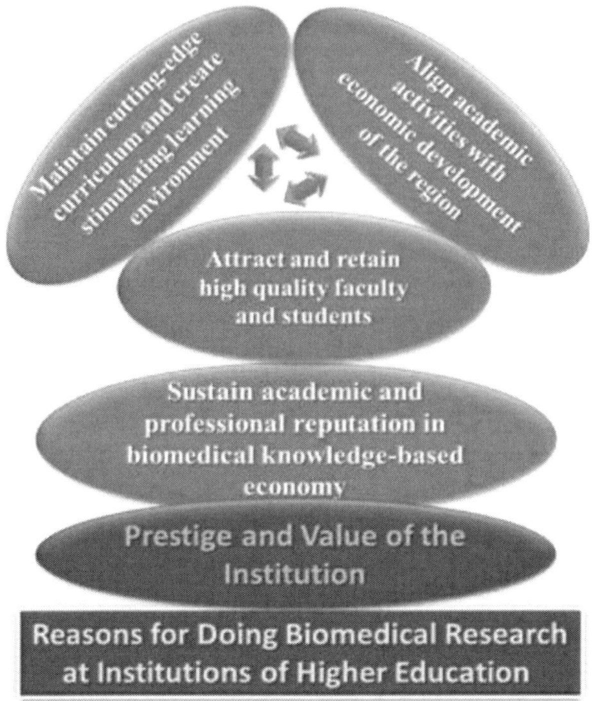

Figure 2-2. Reasons for doing biomedical research at institutions of higher education. As illustrated, biomedical research contributes to the improvement of institutional prestige and value. Top medical schools are exclusively also major biomedical research institutions [21].

Box 2-2. U.S. News & World Report ranking of best medical schools in the United States [22]

U.S. News & World Report is an American news magazine published from Washington, D.C. Along with Time and Newsweek it was for many years a leading news weekly, focusing more than its counterparts on political, economic, health, and education stories. In recent years, it has become particularly known for its ranking system and annual reports on American colleges, graduate schools, medical schools, and hospitals.

U.S. News & World Report 2014 ranking of "Best Medical Schools" in the United States was released in March 2013 (http://grad-schools.usnews.rankingsandreviews.com /best-graduate-schools/top-medical-schools). For the 2014 report, 114 accredited medical schools and accredited osteopathic schools were included in both the research and primary care rankings. The results are based on surveys of these schools during fall 2012 and the beginning part of 2013. It is important to note that multiple osteopathic medical schools fare well in the latest rankings of "Best Medical Schools." In primary care, which is considered one of the two major ranking categories, the Michigan State University College of Osteopathic Medicine in East Lansing stands at No. 11 and the University of North Texas Health Science Center Texas College of Osteopathic Medicine (UNTHSC/TCOM) in Fort Worth at No. 31. TCOM is known for its excellent biomedical research programs, especially the clinical research on osteopathic manipulative medicine.

In addition, the Kentucky College of Osteopathic Medicine at University of Pikeville and the West Virginia School of Osteopathic Medicine (WVSOM) in Lewisburg rank 5th and 9th, respectively, in rural medicine. Multiple osteopathic schools are also highly ranked in certain specialties, including geriatrics and family medicine. Due to the fact that osteopathic medical schools traditionally emphasize primary care, the top ranked research medical schools are mainly allopathic institutions. Nevertheless, as discussed in Section 3.2.2, the osteopathic medicine profession has recently witnessed increased investment in biomedical research and the consequent development of respected biomedical research programs in many osteopathic medical schools in the United States.

Listed below are the top 25 best medical schools in research:

1 Harvard University
2 Stanford University
3 Johns Hopkins University
4 University of California–San Francisco
4 University of Pennsylvania (Perelman)
6 Washington University in St. Louis
7 Yale University
8 Columbia University
8 Duke University
8 University of Chicago (Pritzker)
8 University of Michigan–Ann Arbor
12 University of Washington
13 University of California–Los Angeles (Geffen)
14 Vanderbilt University
15 University of California–San Diego
16 Cornell University (Weill)
16 University of Pittsburgh
18 Baylor College of Medicine
18 Mount Sinai School of Medicine
18 Northwestern University (Feinberg)
21 New York University
22 Emory University
22 University of North Carolina–Chapel Hill
22 University of Texas Southwestern Medical Center
25 Case Western Reserve University

As indicated in **Box 2-2**, multiple osteopathic medical schools are highly-recognized as best medical schools in primary care by the U.S. News & World Report. Osteopathic medicine is a distinct form of medical practice in the United States. As summarized in **Box 2-3** [23], osteopathic medicine provides all of the benefits of modern medicine including prescription drugs, surgery, and the use of technology to diagnose disease and evaluate injury. It also offers the added benefits of hands-on diagnosis and treatment through a system of therapy known as osteopathic manipulative medicine. Osteopathic medicine emphasizes helping each person achieve a high level of wellness by focusing on health promotion and disease prevention [24, 25].

Box 2-3. Osteopathic medical school and osteopathic medical education in brief [23]

- There are currently 29 colleges of osteopathic medicine in the United States, offering instruction at 34 locations in 28 states.
- Today, more than 20 percent of medical students in the United States are being trained to be osteopathic physicians.
- Colleges of osteopathic medicine are graduating more and more students each year. More than 4,200 new osteopathic physicians enter the workforce each year.
- The nation's approximately 63,000 fully licensed osteopathic physicians practice the entire scope of modern medicine, bringing a patient-centered, holistic, hands-on approach to diagnosing and treating illness and injury.
- Osteopathic physicians can choose any specialty, prescribe drugs, perform surgeries, and practice medicine anywhere in the United States.
- Osteopathic physicians bring the additional benefits of osteopathic manipulative techniques to diagnose and treat patients.
- Osteopathic physicians work in partnership with patients to help them achieve a high level of wellness by focusing on health education, injury prevention, and disease prevention.

Traditionally, osteopathic medical schools are largely community-based rather than research university-based. This may account for the relatively smaller scale of biomedical research activities at these schools compared with medical institutions affiliated with research universities [26, 27]. On the other hand, because of the holistic approach it takes, osteopathic medicine offers more opportunities for biomedical research. In this regard, among the research opportunities in osteopathic medicine are investigations of the clinical effectiveness and efficacy and biological basis of OMT as well as other modalities which are relatively unique to osteopathic medicine. These include mind-body medicine, holistic health, exercise, and dietary factors and natural products [28-31] (see Chapters 10 and 11 for more detailed discussion on osteopathic modalities). This has promoted the development of competitive research programs dictated by the institutional strategic research plan at many osteopathic medical schools. Indeed, over the past several years, the osteopathic medicine profession has witnessed increased investment in biomedical research, and the consequent development of respected biomedical research programs in multiple osteopathic medical schools [32-36]. In line with this positive trend, the American Osteopathic Association (AOA) Strategic Plan/Strategic Pathways 2011-2013 (**Box 2-4**) lists research as the No. 2 of the six great strategic pathways that lead to greatness [37]. Strong leaderships with a compelling vision for biomedical research at the level of both professional societies and individual osteopathic medical institutions is essential for ensuring the continued development of basic, translational, and clinical research to advance the osteopathic medicine profession. In this context, vigorously designed biomedical research, including both basic research and clinical trials is warranted to firmly establish the clinical effectiveness and efficacy of OMT and other osteopathic preventive and treatment modalities and the underlying molecular and cellular mechanisms [38-42].

Box 2-4. The American Osteopathic Association's Six Strategic Pathways 2011-2013 [37]

The American Osteopathic Association (AOA) is the national organization representing osteopathic medicine. Its planning cycle produces a triennial strategic plan to guide the work of its bureaus, councils, committees, and staff. The AOA focuses its efforts and resources along "strategic pathways" to become a great association. A strategic pathway is an area of concentration through which the AOA emphasizes the development and enhancement of programs and services across all units of the Association. Strategic pathways provide the flexibility to take advantage of emergent opportunities without losing sight of the strategic direction.

The AOA's 2011-2013 six strategic pathways are:

- Governance
- Research
- Education
- Advocacy
- Teamwork
- Family

Collectively, these strategic pathways represent the means by which the AOA will achieve greatness.

Conclusion and Perspectives

In conclusion, biomedical research largely supported by the United States NIH has contributed greatly to the improvement of public health and social and economic development. Biomedical research is not only vital to public health and the economy, it is also essential to the prestige and value of institutions of higher education, including medical schools. Indeed, the extent to which higher education institutions, including medical colleges are engaged in biomedical research and development activities has a key role in determining the status and quality of these institutions and the contribution which they make to public health, and economic and social development.

While biomedical research has been relatively well-established in many allopathic medical schools, it needs more emphasis in osteopathic medical schools. As such, developing and improving biomedical research have been major challenges that are currently being faced by osteopathic medical schools and also by other schools of health professions affiliated with traditionally non-research-intensive universities. A fundamental understanding of the strategic biomedical research components is a prerequisite for developing biomedical research. The three chapters in Unit II discuss the structural components of strategic biomedical research so as to provide a knowledge framework for developing strategic biomedical research at non-research-intensive institutions of higher education.

References

[1] Allen L. The art of evaluating the impact of medical science. *Bulletin of the World Health Organization* 2010; 88:4-4.

[2] Research!America. 2011 U.S. investment in health research. http://www.research america.org/uploads/healthdollar11.pdf (accessed December 2, 2012).

[3] Research!America. 2010 U.S. investment in health research. http://www.research america.org/uploads/healthdollar10.pdf (accessed December 2, 2012).

[4] NIH. Biennial report of the director of the National Institutes of Health for fiscal years (FYs) 2008 and 2009. http://report.nih.gov/biennialreport/PDF/Ch1_Statement.pdf (accessed December 2, 2012).

[5] Collins FS. Congressional Justification of the NIH fiscal year (FY) 2013 budget request. http://officeofbudget.od.nih.gov/pdfs/FY13/FY2013_Overview.pdf (accessed December 2, 2012).

[6] European Medical Research councils (EMRC*).* A Stronger Biomedical Research for a Better European Future (2011). http://www.esf.org/uploads/media/emrc_wpII.pdf

[7] Epstein JA. Politicizing NIH funding: a bridge to nowhere. *J Clin Invest.* 2011; 121(9): 3362–3363.

[8] Hirschberg R, La Montagne J, Fauci AS. Biomedical research: an integral component of national security. *N Engl J Med.* 2004; 350:2119-21

[9] Statement of Anthony S. Fauci, M.D. Dual Use Research of Concern: Balancing Benefits and Risks. Testimony Before the Committee on Homeland Security and Governmental Affairs, United States Senate (2012). http://www.niaid.nih.gov/about /directors/Documents/DURCfinalClearedTestimony.PDF(accessed December 2, 2012).

[10] Steinbrook R. Biomedical research and biosecurity. *N Engl J Med.* 2005; 353:2212-4.

[11] Enserink M, Malakoff D. Biosecurity. Will flu papers lead to new research oversight? *Science*. 2012; 335:20, 22.

[12] Kaiser J. Biodefense. Proposed biosecurity review plan endorses self-regulation. *Science.* 2007; 316:529.

[13] Fauci AS, Collins FS. Benefits and risks of influenza research: lessons learned. *Science.* 2012; 336:1522-3.

[14] Macilwain C. Science economics: What science is really worth. *Nature.* 2010; 465:682-4.

[14a] Manton KG, Gu XL, Lowrimore G, Ullian A, Tolley HD. NIH funding trajectories and their correlations with US health dynamics from 1950 to 2004. *Proc Natl Acad Sci USA*. 2009; 106:10981-6.

[15] Ehrlich E./United for Medical Research. An economic engine: NIH research, employment, and the future of the medical innovation sector (2011). http://www.unitedformedicalresearch.com/wp-content/uploads/2012/07/UMR_ Economic-Engine.pdf (accessed November 30, 2012).

[16] Battelle. Economic impact of the human genome project (2011). http://www.unitedformedicalresearch.com/wp-content/uploads/2012/08/Economic-Impact-of-the-Human-Genome-PROJECT.PDF (accessed December 2, 2012).

[17] The Economic Impact of Publicly Funded Research Conducted by AAMC-Member Medical Schools and Teaching Hospitals: A Report Prepared for the AAMC by Tripp

Umbach (2011). https://www.aamc.org/download/265994/data/tripp-umbach-research.pdf (accessed December 2, 2012).

[18] Altbach PG, Salmi J. The Road to Academic Excellence: The Making of World-Class Research Universities (Directions in Development) (2011), the World Bank, Washington, D.C.

[19] American Osteopathic Association (AOA). http://www.osteopathic.org/osteopathic-health/about-dos/what-is-a-do/Pages/default.aspx (accessed December 2, 2012).

[20] AMAWIRE. Cost-conscious care: the 7th ACGME core competency? http://www.ama-assn.org/ama/pub/amawire/2011-september-28/2011-september-28-rfs.shtml (accessed December 2, 2012).

[21] Moorman GL, Ford H. Understanding the 7 AOA core competencies. http://www.do-online.org/pdf/cme_sponsconf07moorman.pdf (accessed December 2, 2012).

[22] U.S. News & World Report. 2014 Best Medical Schools. http://grad-schools.usnews.rankingsandreviews.com/best-graduate-schools/top-medical-schools (accessed March 25, 2013).

[23] AACOM. U.S. colleges of osteopathic medicine. http://www.aacom.org/about/colleges/Pages/default.aspx (accessed December 2, 2012).

[24] Wu P, Jonathan Siu J. A brief guide to osteopathic medicine for students by students. http://www.aacom.org/resources/bookstore/Documents/Brief-Guide-to-OME_Final.pdf (accessed December 2, 2012).

[25] AOA. AOA 2012 annual report. http://www.osteopathic.org/inside-aoa/about/leadership/Documents/2012-AnnualReport.pdf (accessed December 2, 2012).

[26] Gevitz N. Researched and demonstrated: inquiry and infrastructure at osteopathic institutions. *J Am Osteopath Assoc.* 2001; 101:174-9.

[27] Howell JD. The paradox of osteopathy. *N Engl J Med.* 1999; 341:1465-8.

[28] Cruser A, Dubin B, Brown SK, Bakken LL, Licciardone JC, Podawiltz AL, Bulik RJ. Biomedical research competencies for osteopathic medical students. *Osteopath Med Prim Care.* 2009; 3:10.

[29] Seffinger MA. Advancing osteopathic medicine through research. *J Am Osteopath Assoc.* 2012; 112:589-90.

[30] Earley BE, Luce H. An introduction to clinical research in osteopathic medicine. *Prim Care.* 2010; 37:49-64.

[31] Licciardone JC, Kearns CM. A new triadic paradigm for osteopathic research in real-world settings. *J Am Osteopath Assoc.* 2012; 112:447-56.

[32] Guillory VJ, Sharp G. Research at US colleges of osteopathic medicine: a decade of growth. *J Am Osteopath Assoc.* 2003; 103:176-81.

[33] Clearfield MB, Smith-Barbaro P, Guillory VJ, Cavalieri TA, Wood DL, Sharp GF, Hahn MB. Research funding at colleges of osteopathic medicine: 15 years of growth. *J Am Osteopath Assoc.* 2007; 107:469-78.

[34] Suminski RR, May LE, Guillory VJ. Research funding at colleges of osteopathic medicine in the United States. *J Am Osteopath Assoc.* 2012; 112:665-72.

[35] Hodge LM. Osteopathic lymphatic pump techniques to enhance immunity and treat pneumonia. *Int J Osteopath Med.* 2012; 15:13-21.

[36] Licciardone JC, Kearns CM, Hodge LM, Bergamini MV. Associations of cytokine concentrations with key osteopathic lesions and clinical outcomes in patients with

nonspecific chronic low back pain: results from the OSTEOPATHIC Trial. *J Am Osteopath Assoc.* 2012; 112:596-605.

[37] AOA. AOA strategic plan 2011-2013 http://www.osteopathic.org/inside-aoa/about /leadership/Documents/aoa-strategic-plan-2011-2013.pdf (accessed December 2, 2012).

[38] Campbell SM, Winkelmann RR, Walkowski S. Osteopathic manipulative treatment: novel application to dermatological disease. *J Clin Aesthet Dermatol.* 2012; 5:24-32.

[39] Franke H. Why reservations remain: a critical reflection about the systematic review and meta-analysis "Osteopathic manipulative treatment for low back pain" by Licciardone et al. *J Bodyw Mov Ther.* 2012; 16:411-5

[40] Snider KT, Snider EJ, Johnson JC, Hagan C, Schoenwald C. Preventative osteopathic manipulative treatment and the elderly nursing home resident: a pilot study. *J Am Osteopath Assoc.* 2012; 112:489-501.

[41] Posadzki P, Ernst E. Osteopathy for musculoskeletal pain patients: a systematic review of randomized controlled trials. *Clin Rheumatol.* 2011; 30:285-91.

[42] Licciardone JC, King HH, Hensel KL, Williams DG. OSTEOPAThic Health outcomes in chronic low back pain: The OSTEOPATHIC Trial. *Osteopath Med Prim Care.* 2008; 2:5.

Self-Assessment Questions

[1] The biomedical research community has long considered research to be vital to the health and wealth of societies, supporting the view attributed to May Lasker, American philanthropist and ardent campaigner for biomedical research: “if you think research is expensive, try disease”. Biomedical research in the United States is an enterprise of:

A. ~5 billion
B. ~10 billion
C. ~40 billion
D. ~140 billion

[2] Which of the following is the biggest support of biomedical research in the United States?

A. Bill & Melinda Gates Foundation
B. Federal government
C. Industry
D. Local government
E. Private universities

[3] Among federal agencies in the United States, which of the following provides the largest support for biomedical research?

A. Center for Disease Control and Prevention
B. Department of Agriculture

C. Food and Drug Administration
D. National Institutes of Health
E. National Science Foundation

[4] Which of the following best describes the U.S. Center for Disease Control and Prevention (CDC)?

A. CDC is responsible for protecting the public health by assuring that medical devices intended for human use are safe and effective.
B. CDC is the directing and coordinating authority on international health within the United Nations' system.
C. CDC is the largest source of funding for medical research in the world, creating hundreds of thousands of high-quality jobs by funding thousands of scientists in universities and research institutions in every state across America and around the globe.
D. CDC works with states and other partners to provide a system of health surveillance to monitor and prevent disease outbreaks.

[5] "Dual Use Research of Concern," or DURC is defined as research that, based on current understanding, can be reasonably anticipated to provide knowledge, information, products, or technologies that could be directly misapplied to pose a significant threat with broad potential consequences to public health and safety, agricultural crops and other plants, animals, the environment, materiel, or national security. Categories of research that should be closely scrutinized for DURC potential include experiments that, for a specific group of agents and toxins, cause the following effects:

A. Conferring resistance to clinically useful prophylactic or therapeutic interventions against an agent or toxin
B. Decrease of the stability or transmissibility of an agent or toxin
C. Diminution of the harmful consequences of an agent or toxin
D. Diminution of the susceptibility of a host population to an agent or toxin
E. Enhancement of the effectiveness of an immunization against an agent or toxin

[6] Which of the following does not rank world universities?

A. The Academic Ranking of World Universities (ARWU)
B. The Higher education Evaluation and Accreditation Council of Taiwan (HEEACT)
C. The U.S. News & World Report
D. Times Higher Education (THE)

[7] There are currently 6 core competencies for MDs. Which of the following has recently been discussed as the possible 7th competency?

A. Cost-conscious care
B. Interpersonal and communication skills

C. Osteopathic manipulation medicine
D. Professionalism

[8] In its Strategic Plan/Strategic Pathways 2011-2013, the American Osteopathic Association (AOA) lists which of the following as the No. 2 of the six great strategic pathways that lead to greatness?

A. Advocacy
B. Education
C. Governance
D. Research
E. Teamwork

Unit II.
Structural Components of Strategic Biomedical Research

Chapter 3

Core Biomedical Research Facilities

Abstract

Developing strong biomedical research programs requires a thorough understanding of the basic structural components of strategic research, among which are core biomedical research facilities. Core research facilities are centralized shared resources that provide access to instruments, technologies, services, as well as expert consultation and other services to scientific investigators. This chapter first describes common institutional core biomedical research facilities and then introduces the core facilities at several prestigious medical institutions in the United States. It also discusses the general principles of management of core research facilities. The chapter is intended to serve as a general reference for constructing core biomedical research facilities at traditionally non-research-intensive institutions of higher education, where teaching is the primary focus. This and the following two chapters provides a knowledge base for developing the physical infrastructures of strategic biomedical research.

Chapter at a Glance

1. Overview

Developing strategic biomedical research requires an understanding of the physical components of biomedical research structure. World-class universities require world-class research facilities. Establishment of competitive biomedical research programs relies on the availability of first-rate biomedical research facilities, which include three essential components:

(1) Institutional core biomedical research facilities
(2) Laboratory animal facilities
(3) Faculty researcher independent laboratories

The institutional success in biomedical research is largely dependent on the availability of advanced core research facilities as well as the synergistic interactions among the three structural components (**Figure 3-1**).

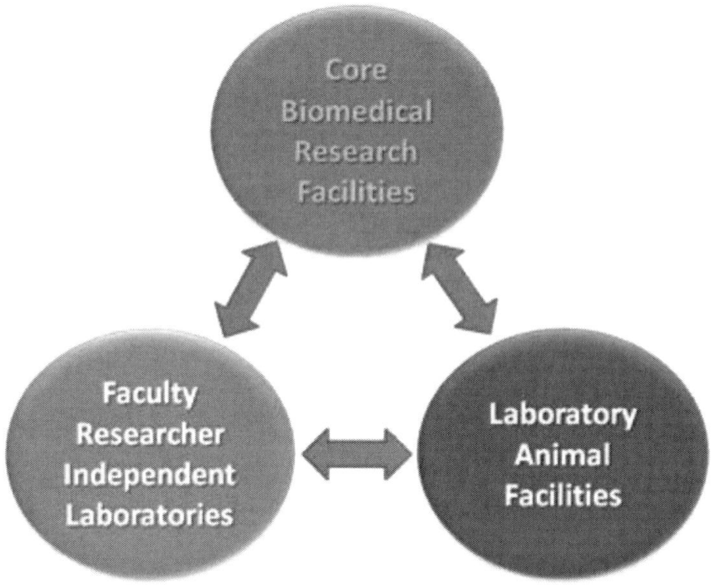

Figure 3-1. Three essential physical components of biomedical research structure. As depicted, the three components do not exist alone, and the synergistic interactions among them are critical for maximizing the institutional research output.

2. Essentiality of Institutional Core Biomedical Research Facilities

2.1. General Considerations

Over the past twenty years, biomedical researchers have increasingly relied on complex instruments that may live in the laboratory of a single researcher or in a core facility. In many areas of basic research, translational science, and clinical studies, access to core facilities has gone from being useful to being essential in the conduct of biomedical research. This is in part because core research facilities provide highly specialized services, equipment, and staff that would otherwise too expensive for a single laboratory or department to provide.

2.2. Defining Core Research Facilities

The United States National Institutes of Health (NIH) defines core research facilities as centralized shared resources that provide access to instruments, technologies, services, as well as expert consultation and other services to scientific investigators. The typical core research facility is a discrete unit within an institution and has dedicated personnel, equipment, and space for operations. In general, core research facilities recover their cost of providing service in the form of user fees that are charged to an investigator's funds, often NIH or other federal grants [1]. Many cores focus on instrumentation, but informatics and biostatistics cores are key resources for translational researchers. The consultation provided by core facilities is often as important as the data because biomedical researchers not trained in a specific technique or field can find it difficult to interpret specialized data without help from core facility experts.

2.3. Essentiality of Core Biomedical Research Facilities

The race is on to develop and effectively use core research facilities to strengthen institutional research enterprises both in the United States and worldwide. In this context, over the past several years, Asian countries, especially China, South Korea, and Singapore have invested tremendously to build world-class core research facilities to strengthen their basic and clinical research. Although the impetus to do so varies across disciplines and institutions, it is becoming increasingly clear that the performance of research in many areas especially biomedicine relies on the availability of complex, expensive technical equipment that often requires operation by dedicated, skilled scientific personnel [2].

Research institutions, including universities, academic medical centers, and independent research institutes, are increasingly realizing the essential role of core biomedical research facilities in advancing their research enterprises as well as the prestige and value of the institutions. Examples of the benefits of establishing institutional advanced core biomedical research facilities include the following [2, 3]:

- Augmenting the researchers' ability to conduct cutting-edge biomedical research
- Making the institution competitive for recruiting and retaining strong faculty members
- Fostering a more collaborative research environment at a university or academic medical center, which is crucial to interdisciplinary research endeavors
- Increasing the institution's competitiveness for external research funding

As noted, having institutional core biomedical research facilities is also essential for research faculty to secure extramural grant funding, especially the grants from NIH. In this regard, the availability and quality of shared core facilities as a key component of the institutional research environment are one of the five criteria that determine the success of applications to NIH grants (**Box 3-1**) [4], as well as other competitive federal and non-federal funds.

Box 3-1. NIH standard criteria and considerations for research project grant (RPG/R01/R03/R15/R21/R34) critiques

1. Significance
2. Investigator(s)
3. Innovation
4. Approach
5. **Environment:**

R01, R03, R21, R34. Will the scientific environment in which the work will be done contribute to the probability of success? Are the institutional support, equipment and other physical resources available to the investigators adequate for the project proposed? Will the project benefit from unique features of the scientific environment, subject populations, or collaborative arrangements?

R15. Are the institutional support, equipment and other physical resources available to the investigators adequate for the project proposed? Does the application demonstrate the likely availability of well-qualified students to participate in the research project? Does the application provide sufficient evidence that students have in the past or are likely to pursue careers in the biomedical or behavioral sciences?

3. Diversity and Complexity of Institutional Core Biomedical Research Facilities

3.1. General Considerations

The diversity and complexity of the core biomedical research facilities vary among institutions and are generally matched by the diversity and complexity of the research institutions in which core research facilities operate. Top medical research institutions, such as Johns Hopkins Medical Institutions and Harvard Medical School are equipped with state-of-the art core research facilities. On the other hand, many non-research-intensive institutions have either no core research facilities or only very limited routine biomedical equipment that is usually used for teaching purpose.

3.2. State-of-the-Art Core Biomedical Research Facilities at Top Medical Institutions

Major research institutions are working to better understand their portfolios of existing core research facilities and to better manage these facilities to fulfill their mission of innovative research. On the other hand, emerging research institutions with plenty of institutional funds are working to create new core facilities in areas such as genomics, proteomics, and bioinformatics to jump-start their research infrastructures and increase their research capacities and competiveness. The core biomedical research facilities at several top medical institutions in the United States are listed in **Table 3-1**, and **Boxes 3-2**, **3-3**, **3-4**, and **3-5**.

Table 3-1. The core research facilities at Johns Hopkins Medical Institutions [5]

Type of Core Research Facilities	Description
Special Animal Core Facilities	SCID Mouse Facility Small Animal Imaging Resource Program Survival Surgery Facilities X-ray Imaging Animal Lab Transgenic Mouse Core Facility
Computing Core Facilities	East Reading Room Interactive Graphics Facility Welch Training Program
NMR Core Facilities	Spectroscopy, High Field (14.1T) Spectroscopy and Imaging
Other Core Facilities	Microarray Core Facility High Throughput Biology Center Drug Development Unit General Clinical Research Center School of Medicine Microscope Facility Flow Cytometry Core Analytic Laboratories Genetic Resources Core Facility Genome Data Base Mass Spectrometry Laboratory, Middle Atlantic MRI Imaging Positron Emission Tomography (PET) Center Protein/Peptide/DNA Facility Biostatistics Center (School of Public Health)

Box 3-2. Core research facilities at Harvard Medical School [6]

Harvard medical school core research facilities are organized as Quad Department Core Facilities, which are listed below.

BCMP SBGrid Computing Core and Software Consortium
BCMP Small Molecule NMR Facility
Cell Biology Conventional Electron Microscopy Facility
Cell Biology Frog Facility
Cell Biology Molecular Electron Microscopy Facility
Cell Biology Taplin Biological Mass Spectrometry Facility
Department of Genetics Biopolymer/Next Gen Sequencing Facility
DF/HCC DNA Resource Core
DF/HCC Rodent Histopathology Core
DF/HCCC Vector Service Center

Genetics Microscopy Facility
Harvard NeuroDiscovery Center Optical Imaging
HMS Microfluidics Core
ICCB-Longwood Screening Facility (ICCB-Longwood)
Microbiology and Immunobiology Analytical Chemistry Facility
Microbiology and Immunobiology BL2+ Imaging Facility
Microbiology and Immunobiology Flow Cytometry Facility
NERCE Biomolecule Production Laboratory
NERCE Microbiology and Animal Resources Core Laboratory
NERCE/Live-cell Imaging Core
NERCE/National Small Molecule Screening and Medicinal Chemistry Core (NSRB)
Neurobiology Imaging Facility
Systems Biology FACS Facility
Systems Biology Machine Shop
Systems Biology Microscopy Facility
The Drosophila RNAi Screening Center (DRSC) at Harvard Medical School
The Image and Data Analysis Core (IDAC)
The Nikon Imaging Center at Harvard Medical School (NIC@HMS)

Box 3-3. Core research facilities at University of Pennsylvania Perelman School of Medicine [7]

The Perelman School of Medicine biomedical research core facilities include 21 research cores that offer a number of services, ranging from molecular profiling to cell sorting to high resolution electron microscopy.

Cell and Developmental Biology Microscopy Core
Cell Center Services Facility
Cell Center Stockroom
Clinical Cell and Vaccine Production Facility
Clinical Research Computing Unit
DNA Sequencing Facility
Electron Microscopy
Flow Cytometry and Cell Sorting Facility
Human Immunology Core
Investigational Drug Service
Molecular Profiling Core
Neurobehavior Testing Core
Next Generation Sequencing Core
Penn Gene Targeting Service
Proteomics Core Facility
Research Instrumentation Shop
Small Animal Imaging Facility
Stem Cell and Xenograft Core
Transgenic and Chimeric Mouse Facility
Translational Core Laboratory
Vector Core

Box 3-4. Core research facilities at Duke University School of Medicine [8]

Duke University offers more than 30 state-of-the-art shared research facilities that are available for use by all Duke investigators. Core facilities are physically housed and administered by Duke research institutes and centers, including the Duke Cancer Institute (DCI), the Institute for Genome Sciences and Policy (IGSP), the Duke Center for Human Genetics (CHG), the Duke Translational Medicine Institute (DTMI), the Duke Center for AIDS Research (CFAR), and the Duke Human Vaccine Institute (DHVI). The main categories of Duke core facilities for biomedical research are listed below. Each category includes multiple co-facilities/resources.

Bioinformatics and Biostatistics
Biospecimen Processing and Storage
Cell Culture and Flow Cytometry
Clinical and Pharmaceutical Services
Crystallography, NMR, and Radiolabeling
DNA, RNA, Molecular, and Genomics Services
HIV/Virology Services
Immune Monitoring and Reconstitution
Microscopy and Imaging
Model Systems and Transgenics
Proteomics and Biochemical Analysis
Regional Biocontainment Laboratory

Box 3-5. Core research facilities at Yale School of Medicine [9]

Yale School of Medicine's core research facilities offer investigators a wide range of state-of-the art instrumentation and technologies for basic, translational, and clinical research. Many of the cores provide training and support services to help investigators make optimal use of the advanced technologies available to them. Highly-trained staff offer assistance at all stages of a project, from writing grant applications and developing research proposals to analyzing, processing, and interpreting the huge amounts of data that are often generated by sophisticated technologies. Listed below are the main categories of core research facilities, and each includes multiple facilities/resources. Each category includes multiple co-facilities/resources.

Animal Technologies
Bioinformatics, Biostatistics, and Computing
Cell- and Tissue-Based Technologies
Drug Discovery
Genomics and Proteomics
Imaging
Metabolic Cluster
Shops
Research Centers (cores at the various research centers)

3.3. Typical, at a Minimum, Components of Core Biomedical Research Facilities

As noted earlier in Chapter 2, developing biomedical research has also been becoming a focus of efforts at many traditionally non-research-intensive teaching institutions that have relatively limited institutional funds to support competitive biomedical research. Leaders at such institutions need to understand that typical institutional core facilities should be in place in order to attract and retain biomedical researchers and jump-start competitive biomedical research. **Table 3-2** lists the typical, at a minimum, core biomedical research facilities.

Table 3-2. Typical, at a minimum, components of core biomedical research facilities

Type of Core Facilities	Description
Cell and Tissue Culture Core	Cell and tissue culture hood, cell culture incubator, refrigerate centrifuge, high speed centrifuge, ultra-speed centrifuge, inverted phase-contrast microscope
Flow Cytometry and Cell Sorting Core	BD Accuri™ C6 flow cytometer (a highly affordable and widely used model; especially suitable for institutions with limited budget)
Histology, Pathology, and Clinical Biochemistry Core	Microtome, staining, blood chemistry analyzer, spectrophotometer
Microscopy and Imaging Core	Regular light microscope, electron microscope, fluorescence microscope
Analytical and Metabolism Core	UV-Vis spectrophotometer, fluorescence spectrophotometer, luminometer, HPLC, Mass spectrometer, electron paramagnetic resonance (EPR) spectroscope, nuclear magnetic resonance (NMR)
Bioinformatics and Biostatistics Core	The objective of the Bioinformatics and Biostatistics Core is to provide advice, training, and assistance in experimental design, data analysis, and statistics, especially for research involving gene and pathway analysis, proteomics, genomics, and clinical and animal studies. Although bioinformatics research can be done via a fee-based service at major research universities, in-house biostatisticians and common statistical analysis softwares must be available to the biomedical research faculty.
Routine Maintenance Core	Water purification system, glassware washing, ice machine, autoclave, gas tank (nitrogen, CO_2, oxygen), liquid nitrogen tank, low-temperature freezer, ultra-low temperature freezer, cold room, dark room

Shared core facilities are indispensible for the success of research institutions. Indeed, as described above, successful and top research institutions always possess advanced core facilities. However, establishment of advanced core facilities is highly costly and often time is not feasible for traditionally non-research-intensive institutions. Nevertheless, without core research facilities, developing biomedical research will be empty talk. Hence, at a minimum, core facilities for non-research-intensive institutions, including osteopathic medical schools should match the typical equipment and facilities listed in Table **3-2**, and include at a minimum the following: typical analytical instruments, microscope and imaging system, low-temperature freezer, high-speed and ultra-speed centrifuges, deionized water system, autoclave/sterile facilities, ice machine, liquid nitrogen tank, carbon dioxide and nitrogen gas tanks, fume hood, tissue culture facilities, and darkroom among many others. These basic shared equipment and facilities are usually purchased, maintained, and upgraded through institutional funds though extramural funds are also available to support core research facilities (see Section 4.1 below and Table 6-2 of Chapter 6 for more information on extramural grants supporting core research facilities).

4. Management of Core Biomedical Research Facilities

4.1. General Considerations

As stated earlier, to conduct high-quality state-of-the-art research, biomedical scientists need access to specialized core research facilities and appropriately trained staff. In this time of economic constraints and increasing research costs, organized and efficient core facilities are essential for researchers who seek to investigate complex biomedical research questions [3]. Each year, NIH invests substantial resources in core research facilities through a wide array of awards [3, 10, 11]. Many core facilities are supported by NIH under the P30 activity code (Table 6-3 of Chapter 6). In FY 2010, NIH funded 493 existing P30 awards and 27 new awards. The total amount awarded was $637 million. This is certainly an underestimate of the support to core facilities because it does not count charges paid by R01 recipients to use the cores, all of the other ways that NIH funds core facilities, or the instrumentation that is awarded to such cores using the S10 activity code. Considering all these, a reasonable estimate is that NIH provides approximately $900 million per year to operate and use core research facilities [3]. Academic institutions also make large monetary investments in core facilities, often by providing matching funds to acquire new instruments, renovate space, and support the salaries of core facility staff members. Such investment can be necessary because many cores cannot cover their actual operating costs by recharges to users of the facilities [3].

4.2. Issues Related to Management of Core Biomedical Research Facilities

The large investment by both the federal government and academic institutions suggests that ways to improve the efficiency of core facilities should be evaluated and implemented when appropriate. In this context, a number of articles have been published over the past years to specifically address the management of core biomedical research facilities to maximize the return on investment and advance the biomedical research enterprise at both research universities and traditionally non-research-intensive institutions [2, 3, 12-16]. Listed below are the commonly encountered issues that need to be effectively addressed in order to increase the efficiency of the core facilities to better serve the research faculty and the institutional biomedical research as a whole.

- Information about core facilities and their resources is either not easily available or is incomplete [3]. Institutional leadership especially at the traditionally non-research-intensive universities may lack an understanding of the essentiality of core research facilities to the institutional biomedical research enterprise. As a result, institutional funds are shifted away from supporting core research facilities.

- Core facility directors and staff members often lack training in the business aspects of core facilities and have uncertain career paths in institutions of higher education. In this regard, the general core operating principles suggested by NIH may serve as a framework for achieving better management [1]. The general best practice for a core research facility is to ensure that:

  (1) The costs of providing service are allowable, allocable, consistently applied and reasonable.
  (2) The rates established to recover these costs are documented and systematically evaluated against actual cost and revised on a regular basis to reflect actual costs.
  (3) These rates are charged to all users on a consistent basis, regardless of funding source(s) and employ a principle of "one service, one rate".

- Institutions are uncertain whether or when large consolidated cores offer better service than small cores. If institutional officials decide to centralize existing cores, they face great difficulty in completing the process [3].
- Government policies and rules for managing and reporting on cores can be unclear and vary substantially depending on the exact terms and conditions of each award. Such variations increase the expense in administering core facilities while adding little benefit to the government [3]. In this context, a recent conference on management and utilization of core research facilities sponsored by NIH has produced insightful suggestions from leading administrators, researchers, and governmental officials in the areas of state-of-the-art core research facilities [17]. These suggestions focused on six key areas, including:

  (1) Implementing standards
  (2) Supporting development and sustainability through planning grants and career training grants
  (3) Supporting both equipment and personnel in core grants
  (4) Promoting full utilization of core facilities
  (5) Supporting collaboration
  (6) Supporting consistency and harmonization

Conclusion and Perspectives

Core research facilities are centralized shared resources that provide access to instruments, technologies, services, as well as expert consultation and other services to scientific investigators. In this time of economic restrains, with the equipment and its maintenance costing more and more each year and the technology growing and expanding all the time, it is very difficult, if not impossible, for individual biomedical researchers to keep pace to maintain competitive research activities [12]. In this context, core biomedical research facilities can offer the latest equipment and services for the newest techniques that individual researchers cannot otherwise afford to possess. However, the benefits of core research facilities can only be maximized with right structural components of the core facilities and under efficient management. It is important that institutional leadership, especially in the traditionally non-research-intensive universities, fully appreciates the essentiality of core research facilities to institutional biomedical research enterprise. Policies for encouragement of core facilities are important considerations in planning the budgets of university departments. Researchers should be offered incentives to use core research facilities rather than purchase their own expensive equipment [12]. Core research facility managers need to be well trained and skilled operators and be prepared to keep the core research facilities in the best possible condition and for the best possible outcomes.

References

[1] NIH: Core facilities FAQ. http://cancer.osu.edu/.../NIH%20-%20Core%20Facilities%20FAQs.pdf (accessed on December 3, 2012).

[2] Haley R. A framework for managing core facilities within the research enterprise. *J Biomol Tech.* 2009; 20:226–230.

[3] Farber GK, Weiss L. Core facilities: maximizing the return on investment. *Sci Transl Med.* 2011; 3: 95cm21.

[4] NIH. Definitions of criteria and considerations for research project grant (RPG/R01/R03/R15/R21/R34) critiques. http://grants.nih.gov/grants/peer/critiques/rpg.htm (accessed on December 3, 2012).

[5] Johns Hopkins medical Institutions. CORE research facilities. http://www.hopkinsmedicine.org/research/ora/handbook/corefacilities.html (accessed on December 3, 2012).

[6] Harvard Medical School. HMS core facilities. http://hms.harvard.edu/content/hms-core-facilities (accessed on December 3, 2012).

[7] University of Pennsylvania Perelman School of Medicine. Biomedical research core facilities. http://www.med.upenn.edu/cores/ (accessed on December 3, 2012).

[8] Duke University School of Medicine. Core research facilities. http://medschool.duke.edu/research/core-research-facilities (accessed on December 3, 2012).

[9] Yale School of Medicine. Core research facilities. http://medicine.yale.edu/cores/index.aspx (accessed on December 3, 2012).

[10] Rosenblum D. Access to core facilities and other research resources provided by the Clinical and Translational Science Awards. *Clin Transl Sci.* 2012; 5:78-82.

[11] NIH. Core facilities: accomplishments and plans (2011). http://www.abrf.org/Other/Farber_Flow_cores.pdf (accessed on December 3, 2012).

[12] DeMaggio S. Running and setting up a confocal microscope core facility. *Methods Cell Biol.* 2002; 70:475-85.

[13] Lewitter F, Rebhan M. Establishing a successful bioinformatics core facility team. *PLoS Comput Biol.* 2009; 5:e1000368.

[14] Lilley KS, Deery MJ, Gatto L. Challenges for proteomics core facilities. *Proteomics.* 201; 11:1017-25.

[15] Haley R. Institutional management of core facilities during challenging financial times. *J Biomol Tech.* 2011; 22:127-30.

[16] Slaughter C. A bright but demanding future for core facilities. *J Biomol Tech.* 2005; 16:167-9.
[17] NIH. Final workshop report on the efficient management and utilization of core facilities-sponsored by the National Center for Research Resources and The Office of Extramural Research, Office of the Director, National Institutes of Health , July 14 - 15, 2009. http://dpcpsi.nih.gov/orip/documents/final_workshop_report_july09%20% 281% 29.pdf (accessed on December 3, 2012).

Self-Assessment Questions

[1] Developing strong biomedical research programs relies on a profound understanding of the basic structural components of strategic research, among which are core biomedical research facilities. Which of the following best describes core biomedical research facilities?

A. In general, core research facilities recover their costs of providing service in the form of user fees that are charged to the State government.
B. Many cores focus on instrumentation, because informatics and biostatistics cores are not useful for translational researchers.
C. The term core research facilities is defined by the U.S. NIH as centralized shared resources that provide access to only major, expensive instruments, to scientific investigators.
D. The typical core research facility is a discrete unit within an institution and has dedicated personnel, equipment, and space for operations.
E. Institutional core research facilities should not be made available to investigators not affiliated with the institution.

[2] Core research facilities are supported by which of the following U.S. National Institutes of Health (NIH) grant mechanisms?

A. P30
B. R15
C. R21
D. R25

[3] In a recent meeting with the university senior administrators on establishing institutional core biomedical research facilities, Professor C.F. lists several benefits that core research facilities may provide. Which of the following is most unlikely to be on the list?

A. Augmenting the researchers' ability to conduct cutting-edge biomedical research
B. Fostering a more collaborative research environment at the university
C. Increasing the institution's competitiveness for external research funding
D. Making the institution competitive for recruiting and retaining academic talent
E. Serving as a rationale for raising tuition for graduate programs

[4] The general best practice for a core biomedical research facility is to ensure that:

A. The costs of providing service should be high enough to make good profits.
B. The rates established to recover the costs are documented and systematically evaluated against actual costs and revised on a regular basis to reflect actual costs.
C. The rates should be determined by the National Institutes of Health (NIH) for research supported through NIH grants.
D. The rates that are charged to users are dependent on the funding source(s) and the principle of "one service, one rate" is not suitable for the management of core research facilities.

[5] During a summer research in Professor T.P.'s lab, a first year medical student needs an instrument to complete his experiment. This instrument is costly and only available in the university's Analytical and Imaging Core. Which of the following is most likely the equipment that the medical student needs?

A. A carbon dioxide incubator
B. A pH meter
C. A water bath
D. An electron microscope

Chapter 4

Laboratory Animal Facilities

Abstract

Laboratory animal research plays a vital role in nearly every major medical advance in the last century leading to the development of treatments and cures for numerous human diseases. As such, humans owe a debt of gratitude to the part laboratory animals play in saving the lives of millions of people throughout the world. It is for this and many other reasons that laboratory animals receive humane treatment. This chapter introduces the definitions and basic concepts related to laboratory animal research and focuses on discussing the principles of design and construction of animal facilities as well as the facility accreditation. The policies and guidelines on humane care and use of laboratory animals are covered in Chapter 7.

Chapter at a Glance

1. Overview

Laboratory animal facilities are an essential physical component of biomedical research. While not all biomedical research involves the use of animals on a daily basis, animals are indispensible in many areas of biomedical research. In the past century, most medical discoveries were possible because of animal research. Every person in the United States and throughout the whole world has benefited from the results of biomedical research involving the use of laboratory animals. To build on the great progress we have made in understanding and treating diseases we need to continue the biomedical research using animals. Through biomedical research using laboratory animals, scientists strive:

- to better understand the causes of diseases,
- to develop new and more effective drugs, vaccines, or procedures to prevent or treat diseases, and
- to test the safety of products that we use every day of our lives [1, 2].

As illustrated in **Figure 4-1**, the biomedical achievements derived from animal research not only benefit humans, but also non-human creatures, including our pets, wildlife, and other animals [1]. This chapter first defines such terms as laboratory animal, animal use, and animal facilities and then discusses general guidelines for design and accreditation of institutional laboratory animal facilities.

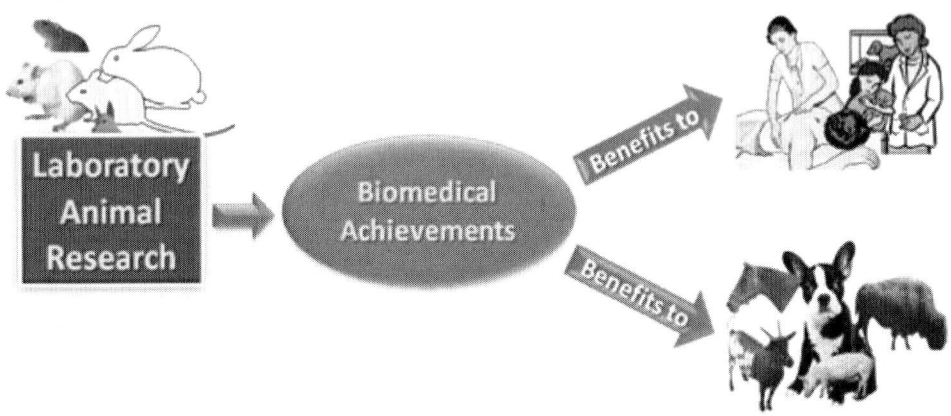

Figure 4-1. Biomedical research involving laboratory animals benefits humans as well as non-human creatures.

2. Definitions

2.1. Laboratory Animal

The Guide for the Care and Use of Laboratory Animals (currently in its 8th edition by the United States National Research Council of the National Academies; also known as the 2011 Guide; **Box 4-1**) is an internationally accepted primary reference on animal care and use, and its use is required in the United States by the Public Health Service Policy [3] (see Chapter 7 for more discussion).

Box 4-1. The Guide for the Care and Use of Laboratory Animals (the Guide)

The Guide is an internationally accepted primary reference on animal care and use, and its use is required in the United States by the Public Health Service Policy. It was first published in 1963, under the title Guide for Laboratory Animal Facilities and Care, and was revised in 1965, 1968, 1972, 1978, 1985, and 1996. More than 550,000 copies have been printed since its first publication.

The most current edition is the 8th edition, which is available at http://www.nap.edu/catalog.php?record_id=12910

The eighth edition of the Guide (known as the 2011 Guide) is divided into five chapters and four appendices:

- Chapter 1: Key concepts
- Chapter 2: Animal care and use program
- Chapter 3: Environment, housing, and management
- Chapter 4: Veterinary care
- Chapter 5: Physical plant
- Appendix A: Additional selected references
- Appendix B: U.S. government principles for the utilization and care of vertebrate animals used in testing, research, and training
- Appendix C: Statement of task
- Appendix D: About the authors

According to the 2011 Guide, the term laboratory animal (also referred to as animal for simplicity) is defined broadly as any vertebrate animal (e.g., traditional laboratory animals, agricultural animals, wildlife, and aquatic species) produced for or used in research, testing, or teaching [3]. The 2011 Guide also defines the term animal use and provides guidelines for designing laboratory animal facilities (see below). The Office of Laboratory Animal Welfare (OLAW) of the United States National Institutes of Health defines laboratory animal as any live, vertebrate animal used or intended for use in research, research training, experimentations, or biological testing, or for related purpose [4]. The Association for Assessment and Accreditation of Laboratory Animal Care International (AAALAC; formerly known as American Association for Accreditation of Laboratory Animal Care) defines laboratory animal as any live vertebrate animal (and any other animal designated by applicable legislation) used or intended for use in research, testing, or teaching. For accreditation purposes, the definition is not limited by the type of housing enclosure (e.g., cage, pen, paddock, pasture, tank, raceway, etc.). All such animals owned by the accreditable unit are to be included in the Program Description (see Section 4.2.3 and **Table 4-3** for definition of Program Description) and will be evaluated by AAALAC [5]. The term animal is also defined in the European Convention for the Protection of Vertebrate Animals Used for Experimental and Other Scientific Purposes (ETS 123) [6]. As summarized in **Table 4-1**, the various definitions of laboratory animal are essentially the same.

Table 4-1. Defining laboratory animal

ORGANIZATION	DEFINITION
The Guide for the Care and Use of Laboratory Animals (8th edition, 2011) [3]	Any vertebrate animal (e.g., traditional laboratory animals, agricultural animals, wildlife, and aquatic species) produced for or used in research, testing, or teaching
Public Health Service Policy on Humane Care and Use of Laboratory Animals (OLAW of NIH 2002) [4]	Any live, vertebrate animal used or intended for use in research, research training, experimentations, or biological testing or for related purpose
AAALAC [5]	Any live vertebrate animal (and any other animal designated by applicable legislation) used or intended for use in research, testing, or teaching
The European Convention for the Protection of Vertebrate Animals Used for Experimental and Other Scientific Purposes (ETS 123) [6]	Animal, unless otherwise qualified, means any live non-human vertebrate, including free-living and/or reproducing larval forms, but excluding other fetal or embryonic forms

2.2. Animal Use

Animal use is an important term in animal research that has specific meaning. The 2011 Guide for the Care and Use of Laboratory Animals (8th edition) defines the term "animal use" as the proper care, use, and humane treatment of laboratory animals produced for or used in research, testing, or teaching [3].

2.3. Animal Facilities

According to the Public Health Service Policy on Humane Care and Use of Laboratory Animals (released by OLAW of NIH, 2002) [4], the term animal facility refers to any and all buildings, rooms, areas, enclosures, or vehicles, including satellite facilities, used for animal confinement, transport, maintenance, breeding, or experiments inclusive of surgical manipulation. A satellite facility is any containment outside of a core facility or centrally designated or managed area in which animals are housed for more than 24 hours.

3. Design and Construction of Animal Facilities

As mentioned earlier, the 2011 Guide for the Care and Use of Laboratory Animals discusses helpful hints for planning and designing animal facilities and also provides construction guidelines as well as security and access control [3]. Similarly, the Division of Technical Resources at the Office of Research Facilities of the National Institutes of Health also released the 2008 National Institutes of Health (NIH) Design Requirements Manual (DRM) for Biomedical Laboratories and Animal Research Facilities, formerly called the NIH Design Policy and Guidelines [7]. It is the only detailed design requirements and guidance manual for biomedical research laboratory and animal research facilities in the United States.

Compliance to the NIH DRM for Biomedical Laboratories and Animal Research Facilities, which promulgates minimum performance design standards for NIH owned and leased new buildings and renovated facilities, ensures that those facilities will be of the highest quality to support biomedical research. Although the manual is for NIH facilities, it provides a very useful reference for designing biomedical research laboratories and laboratory animal facilities at institutions of higher education. Readers are advised to refer to these two documents for detailed guidelines on design and construction of laboratory animal facilities. This section is intended to provide a brief description of the general principles on designing laboratory animal facilities.

3.1. General Principles of Animal Facility Design and Construction

Outlined below are the general principles of design and construction of animal facilities for biomedical research [3, 7].

(1) A well-planned, well-designed, well-constructed, properly maintained and managed animal facility is an important element of humane animal care and use as it facilitates efficient, economical, and safe operation.

(2) As stated in the 2011 Guide [3], the design and size of an animal facility depend on several factors including: (a) the scope of institutional biomedical research activities, (b) the animal species to be housed, (c) the physical relationship to the rest of the institution, and (d) the geographic location among others. **Box 4-2** outlines the space requirement for animal facilities. **Box 4-3** describes the animal species used in biomedical research.

Box 4-2. Space requirement for animal facilities: factors to consider [7]

- The space requirements for animal facilities are among the most challenging issues and may vary greatly.
- Space requirements are dependent on various aspects, including (i) the specific use of the facility, (ii) type and density of animals housed, (iii) type of caging and racking systems, (iv) number of investigators utilizing the facility, and (v) operational methodologies of the facility.
- Each proposed facility shall require careful analysis by the architects/engineers and consultation with users to determine adequate space requirements.
- The space requirements for an animal facility shall consider (i) the total animal population, (ii) number of species, (iii) isolation requirements, (iv) number of animals per room, and (v) number of investigators and research projects anticipated.
- The assignment of support space is based on protocol, equipment, and process and can be determined only on the basis of an evaluation of the specific project program of the facility users.
- Application of these space criteria requires the architects/engineers to analyze functional requirements in light of specific project needs.

Box 4-3. Animal species used in biomedical research: a glance [1, 2]

With regard to the animal species used in biomedical research in the United States as well as in most research institutions throughout the world, more than 90% of them are mice, rats, and other rodents. The rest include dogs, cats, and primates. In most cases, these animals are specifically bred for research purposes and are purchased from animal breeders.

(3) Effective planning and design should include input from personnel experienced with animal facility design, engineering, and operation, as well as from representative users of the proposed facility.

(4) An animal facility should be designed and constructed in accord with all applicable building codes.

(5) Biomedical research is heavily dependent on animal research to create animal models for the study of human disease processes. Fluctuations in the animal species of choice may vary from time to time, and for this reason the facilities must be capable of meeting ever changing animal research requirements. The design of these facilities must be flexible and adaptable. **Box 4-4** outlines the planning criteria for animal facilities as described by NIH.

Box 4-4. Planning criteria for animal facilities [7]

The animal research facility should be designed to house animals in an appropriate species specific environment that meets or exceeds all applicable policies, guidelines and regulations as outlined in the Guide for the Care and Use of Laboratory Animals (2011 Guide), PHS policy and animal welfare regulations. In addition, facilities must meet the minimum requirements to be accredited by AAALAC. The ideal facility will:

- Meet projected holding and programmatic requirements, while providing for expansion and flexibility in space utilization
- Provide for efficiency of management through innovation and flexible design
- Be cost effective in design, construction, operation, and maintenance
- Utilize innovative design and construction to minimize future energy, maintenance, labor, and expansion costs
- Provide an ergonomic and user friendly work environment
- Include redundancies
- Have capacity for increased holding space

3.2. Minimum Functional Areas

Professional judgment should be exercised in the development of a practical, functional, and efficient physical plant for animal care and use. The size, nature, and intensity of an institutional biomedical research will determine the specific facility and support functions needed. Typically, functional areas of an animal facility in a traditionally non-research-intensive institution should include the space for the following [3]:

- Animal housing, care, and sanitation
- Receipt, quarantine, separation, and/or rederivation of animals
- Separation of species or isolation of individual projects when necessary
- Storage

3.3. Additional Functional Areas

As outlined in the 2011 Guide, most multipurpose animal facilities may also include the following additional functional areas [3]:

- Specialized laboratories or space contiguous with or near animal housing areas for such activities as surgery, intensive care, necropsy, irradiation, preparation of special diets, experimental procedures, behavioral testing, imaging, clinical treatment, and diagnostic laboratory procedures
- Containment facilities or equipment, if hazardous biologic, physical, or chemical agents are to be used
- Barrier facilities for housing of specific pathogen-free rodents, especially valuable genetically modified animals, or irreplaceable animal models
- Receiving and storage areas for food, bedding, pharmaceuticals, biologics, and supplies
- Space for washing and sterilizing equipment and supplies and, depending on the volume of work, machines for washing cages, bottles, glassware, racks, and waste cans; a utility sink; a sterilizer for equipment, food, and bedding; and separate areas for holding soiled and clean equipment
- Space for storing wastes before incineration or removal
- Space for cold storage or disposal of carcasses
- Space for administrative and supervisory personnel, including space for staff training and education
- Showers, sinks, lockers, toilets, and break areas for personnel
- Security features, such as card-key systems, electronic surveillance, and alarms
- Areas for maintenance and repair

4. Accreditation of Animal Facilities

4.1. General Considerations

According to the 2002 Public Health Service Policy on Humane Care and Use of Laboratory Animals, each institution must assure that its animal facilities are in one of the following two categories [3]:

- Category 1- Accredited by AAALAC. All of the institution's programs and facilities (including satellite facilities) for activities involving animals have been evaluated and accredited by AAALAC, which is the only accrediting body recognized by the United States Public Health Service (PHS). All of the institution's programs and facilities (including satellite facilities) for activities involving animals have also been evaluated by the Institutional Animal Care and Use Committee (IACUC) (see Chapter 7 for detailed discussion of IACUC) and will be reevaluated by the IACUC at least once every six months, and reports prepared in accordance with the requirements of this Policy.
- Category 2- Evaluated by the institution. All of the institution's programs and facilities (including satellite facilities) for activities involving animals have been evaluated by the IACUC and will be reevaluated by the IACUC at least once every six months, and reports prepared in accordance with requirements of this Policy.

According to the above 2002 PHS Policy, PHS-sponsored research can be carried out in animal facilities that are either accredited by AAALAC or evaluated by the institution. In other words, AAALAC accreditation is not mandatory for institutions to receive PHS grants to support biomedical research involving the use of laboratory animals at the institution. However, as discussed next, AAALAC accreditation comes with many benefits. It is important to also note that as of the 2002 revision of the PHS Policy, the only accrediting body recognized by PHS is AAALAC. In this context, accreditation by non-AAALAC accreditation bodies is not advised.

4.2. Accreditation by AAALAC

4.2.1. Benefits of AAALAC Accreditation

AAALAC accreditation benefits an institution and the laboratory animals in its care in numerous ways. Each time a new institution becomes AAALAC accredited, it helps to raise the global benchmark for animal well-being in scientific research. **Table 4-3** lists seven major benefits of AAALAC accreditation [8].

Table 4-3. Seven major benefits of AAALAC accreditation [8]

BENEFIT	DESCRIPTION
It Represents Quality	Organizations and companies look for ways to communicate their commitment to excellence. In the scientific community, AAALAC accreditation shows that an institution is serious about setting, achieving and maintaining high standards for animal care and use and committed to animal welfare in science. AAALAC offers the only international accreditation for animal care and use programs, and it has become recognized around the world as a sign of quality and good science.
It Promotes Scientific Validity	When research involves animals, reliable scientific results depend on superior animal care. AAALAC accreditation engages scientists, veterinarians, managers, and administrators in an independent, rigorous assessment of their institution's animal program- an assessment that ultimately results in improved animal welfare and better research practices and outcomes.

Benefit	Description
It Provides Assurance in a Global Marketplace	Today it's common for research institutions to partner or contract with other research entities around the world. Because laws and regulations related to animal research vary widely from country to country, AAALAC accreditation can be used worldwide as a way to gauge the quality of a particular program, harmonize animal care and use practices, and provide assurance to diverse stakeholders.
It's a Recruiting Tool	AAALAC accredited institutions can use their accreditation as a recruiting tool to attract the best and brightest researchers and professors. Talented professionals look for high quality programs to support their research. Accreditation assures potential employees that the institution is dedicated to achieving the highest standards for animal care and use.
It Demonstrates Accountability	In today's world, companies and organizations are held to very high levels of accountability by their own constituents and the general public. Although animal research is a controversial issue for some, most people support biomedical research if it's conducted in a humane manner. Accreditation through AAALAC is voluntary and demonstrates a willingness to go above and beyond the minimums required by law. It tells the public that the institution is committed to the responsible care and use of animals in science.
It Provides a Confidential Peer-Review	AAALAC accreditation requires an institution to first perform its own self-evaluation (an extremely valuable exercise for any institution to undertake). Next, a team of highly qualified AAALAC representatives provides an in-depth, confidential, on-site evaluation of the institution's animal care and use program. This independent peer-review ensures that the institution's program is meeting AAALAC standards.
It Stimulates Continuous Improvement	When an institution participates in the AAALAC accreditation program, it's committing to a process that stimulates continuous improvement. Earning and maintaining accreditation keeps an institution aware of, and engaged in, current best practices. Accreditation is a true commitment to humane animal care and use and shows the world that an institution is serious about ensuring animal well-being and conducting good science.

4.2.2. Three Primary Standards Used by AAALAC for Accreditation

AAALAC's Council on Accreditation uses three primary standards to evaluate animal care and use programs [9]. They are listed below.

- The 8th Edition of the Guide for the Care and Use of Laboratory Animals (the 2011 Guide), NRC, 2011 [3]
- The Guide for the Care and Use of Agricultural Animals in Research and Teaching (the Ag Guide), FASS, 2010 [10]
- The European Convention for the Protection of Vertebrate Animals Used for Experimental and Other Scientific Purposes, Council of Europe (ETS 123), 1986 (Appendix A in 2006) [6]

4.2.3. Steps of AAALAC Accreditation

The AAALAC accreditation program evaluates organizations that use animals in research, teaching, or testing. Those that meet or exceed AAALAC standards are awarded accreditation. **Table 4-4** summarizes the major steps of AAALAC accreditation [11]. The detailed accreditation process is described on the website of AAALAC (www.aaalac.org).

Table 4-4. AAALAC accreditation steps

STEP	DESCRIPTION
Internal Review	The accreditation process begins with an extensive internal review conducted by the institution applying for accreditation. During this review, the institution creates a comprehensive document called a "Program Description" which describes all aspects of the animal care and use program (policies, animal housing and management, veterinary care, and facilities). The Program Description is then submitted to AAALAC.
AAALAC Review and On-Site Assessment	AAALAC evaluators (members of AAALAC's Council on Accreditation) review the Program Description and conduct their own comprehensive on-site assessment. The site visitors' report is then reviewed by the entire Council on Accreditation and accreditation status is determined. If deficiencies are found, they are outlined in a letter and the institution is given a period of time to correct them. Once the deficiencies are corrected, accreditation is awarded. The entire process is completely confidential. Currently more than 850 organizations in 36 countries have earned AAALAC accreditation.
AAALAC Re-Evaluation	After an institution earns accreditation, it must be re-evaluated every three years in order to maintain its accredited status.

Conclusion and Perspectives

Animal facilities are a critical physical component required for biomedical research at both research universities and traditionally non-research-intensive institutions deemed to develop biomedical research enterprise. In house animal facilities are also essential for obtaining extramural funding, especially NIH grants because successful biomedical research grant applications often times (if not always) involve animal experiments. Animal studies are also frequently carried out to lay a foundation for subsequent clinical research. In the United States, animal research facilities requiring the use of animals must conform to the 2011 Guide to be accredited by AAALAC. For accreditation the environment within the facilities must provide for the health, safety, comfort, and well being of the animals and staff. During planning, it is crucial to identify the variety of species to be housed in the facility over time; the temperature and humidity range that each species can tolerate; and the degree of flexibility and adaptability required to accommodate different species. Vibration stability, noise damping, diurnal lighting, prep space, surface finishes, sealing and caulking, and ventilation are other critical considerations in designing an animal research facility. Since animal facilities present some of the most challenging pest management concerns, the NIH's integrated pest management program [12] may be incorporated into the design.

References

[1] AAALAC. The use of animals in biomedical research: improving human and animal health. http://www.aalas.org/pdf/improve_human_animal_health.pdf (accessed on December 4, 2012).

[2] AAALAC. Use of animals in biomedical research: understanding the issues. http://www.aalas.org/pdf/08-00007.pdf (accessed on December 4, 2012).

[3] National Research Council/National Academies Press. Guide for the Care and Use of Laboratory Animals, 8th ed (2011), Washington D.C. http://www.nap.edu/catalog.php?record_id=12910 (accessed on December 4, 2012).

[4] OLAW/NIH. Public Health Service Policy on Humane Care and Use of Laboratory Animals (2002). http://grants.nih.gov/grants/olaw/references/phspolicylabanimals.pdf (accessed on December 4, 2012).

[5] AAALAC. Definition of laboratory animals. http://www.aaalac.org/accreditation/positionstatements.cfm#labanimals (accessed on December 4, 2012).

[6] European Convention for the Protection of Vertebrate. Animals used for Experimental and Other Scientific Purposes. http://www.aaalac.org/about/ETS123.pdf (accessed on December 4, 2012). Appendix A: http://www.aaalac.org/about/AppA-ETS123.pdf (accessed on December 4, 2012). Appendix B: http://www.aaalac.org/about/AppB-ETS123.pdf (accessed on December 4, 2012).

[7] Office of Research Facilities/NIH. The 2008 National Institutes of Health (NIH) Design Requirements Manual for Biomedical Laboratories and Animal Research Facilities (DRM). http://orf.od.nih.gov/PoliciesAndGuidelines/BiomedicalandAnimal Research FacilitiesDesignPoliciesandGuidelines/DRMHTMLver/DesignRequirementsManualTableofContents.htm (accessed on December 4, 2012).

[8] AAALAC. Benefits of AAALAC international accreditation http://www.aaalac.org/accreditation/benefits.cfm (accessed on December 4, 2012).

[9] AAALAC. AAALAC's *three* primary standards. http://www.aaalac.org/about/guidelines.cfm (accessed on December 4, 2012).

[10] The Federation of Animal Science Societies (FASS). Guide for the Care and Use of Agricultural Animals in Research and Teaching. http://www.fass.org/docs/agguide3rd/Ag_Guide_3rd_ed.pdf (accessed on December 4, 2012).

[11] AAALAC. Step to earning accreditation. http://www.aaalac.org/accreditation/steps.cfm (accessed on December 4, 2012).

[12] NIH. The integrated pest management program. http://www.ors.od.nih.gov/sr/dohs/HealthAndSafety/IPM/Pages/pest_management.aspx (accessed on December 4, 2012).

Self-Assessment Questions

[1] Laboratory animal research plays a vital role in nearly every major medical advance in the last century, leading to the development of treatments and cures for numerous human diseases. As such, humans owe a debt of gratitude to the part laboratory animals play in saving the lives of millions of people throughout the world. It is for this and many other reasons that laboratory animals receive humane treatment. Which of the following is an internationally accepted primary reference on animal care and use, and its use is required in the United States by the Public Health Service Policy.

A. The 2008 U.S. National Institutes of Health (NIH) Design Requirements Manual for Biomedical Laboratories and Animal Research Facilities (DRM)

B. The European Convention for the Protection of Vertebrate Animals Used for Experimental and Other Scientific Purposes (ETS 123)
C. The Guide for the Care and Use of Laboratory Animals (8th Edition, 2011)
D. The U.S. FDA Code of Federal Regulations 21CFR58: Good Laboratory Practice for Preclinical Studies

[2] According to the Guide for the Care and Use of Laboratory Animals (8th edition, 2011), which of the following are not classified as laboratory animals?

A. Dogs
B. Fish
C. Mice
D. Monkeys
E. Rats

[3] The term animal facility is defined in which of the following documents?

A. The American Osteopathic Association (AOA) Strategic Plan/Strategic Pathways 2011-2013
B. The ICH Good Clinical Practice guidance (1996)
C. The Public Health Service Policy on Humane Care and Use of Laboratory Animals (released by OLAW of the U.S. NIH, 2002)
D. The U.S. Federal Research Misconduct (2002)

[4] With regard to the animal species used in biomedical research in the United States as well as in most research institutions throughout the world, more than 90% of them are:

A. Dogs and cats
B. Fish and turtles
C. Flies
D. Mice, rats, and other rodents
E. Non-human primates

[5] According to the U.S. 2002 Public Health Service (PHS) Policy on Humane Care and Use of Laboratory Animals, which of the following is the only accrediting body recognized by the PHS Policy?

A. Association for Assessment and Accreditation of Laboratory Animal Care (AAALAC)
B. Institutional Animal Care and Use Committee (IACUC)
C. Institutional Review Board (IRB)
D. Office of Laboratory Animal Welfare (OLAW)

[6] A medical student is informed by a senior researcher that he is not allowed to conduct animal research in the animal facility of the university because the facility has not been accredited by the Association for Assessment and Accreditation of Laboratory Animal

Care (AAALAC). The medical student argues that conducting animal research in a non-AAALAC-accredited animal facility does not necessarily violate the U.S. Public Health Service (PHS) Policy on Humane Care and Use of Laboratory Animals. The medical student is right because:

A. The animal facility has been accredited by a local accreditation body.
B. The animal facility has been evaluated by the Institutional Animal Care and Use Committee (IACUC) and will be reevaluated by the IACUC at least once every six months.
C. The animal facility has been operated on a full time basis by at least five staff members.
D. The animal facility is directed by a veterinarian.

[7] The Association for Assessment and Accreditation of Laboratory Animal Care (AAALAC)'s Council on Accreditation uses three primary standards to evaluate animal care and use programs. Which of the following is one of the three standards?

A. The 2008 U.S. National Institutes of Health (NIH) Design Requirements Manual for Biomedical Laboratories and Animal Research Facilities (DRM)
B. The ICH Good Clinical Practice guidance (1996)
C. The 8th Edition of the Guide for the Care and Use of Laboratory Animals, NRC, 2011
D. The U.S. FDA Code of Federal Regulations 21CFR58: Good Laboratory Practice for Preclinical Studies

[8] The Office of Laboratory Animal Welfare (OLAW) is part of:

A. The U.S. Department of Agriculture (USDA)
B. The U.S. Food and Drug Administration (FDA)
C. The U.S. National Institutes of Health (NIH)
D. The World Health Organization (WHO)

Chapter 5

Faculty Researcher Independent Laboratories

Abstract

The research success of academic institutions is largely dependent on the quality and productivity of individual investigators. At institutions of higher education, faculty researcher independent laboratories are probably one of the most crucial structural components of institutional biomedical research. The ready availability and high quality of the research laboratories are indispensible for institutional research success. This chapter provides an overview of the definition, classification, and general features of biomedical research laboratories and introduces the principles of planning and designing of the research laboratories. The chapter provides recommendations on faculty researcher laboratory and start fund allocation. It also introduces the concept of modular laboratories as a cost-efficient strategy for establishing biomedical research laboratories at institutions that lack funds for building conventional research buildings.

Chapter at a Glance

1. Overview

The investigator independent research laboratories to an institution's biomedical research enterprise are just like cells to a tissue. Without functional cells, the tissue or organ cannot perform its desired physiological functions. Similarly, the success of a research institution heavily depends on the research productivity and quality of investigators' independent research laboratories. This chapter first outlines the basic features of biomedical research laboratories and then describes the general guidelines on planning and design of investigators' independent biomedical research laboratories. The chapter focuses on the guidelines in the 2008 National Institutes of Health (NIH) Design Requirements Manual for Biomedical Laboratories and Animal Research Facilities. The recommendations on the allocation of individual research laboratory space and start fund for newly recruited biomedical faculty researchers are provided toward the end of the chapter.

2. Definition and Classification of Biomedical Research Laboratories

A research laboratory is a workplace where scientific research is conducted. Likewise, a biomedical research laboratory is a workplace for the conducting of biomedical research, including both basic and translational research as well as certain parts of clinical research. At institutions of higher education, a typical biomedical research laboratory is run by a principal investigator/faculty researcher, whose name is attached to the laboratory. The principal investigator is responsible for choosing research topics, raising funds, juggling budgets, and managing postdoctoral fellows and graduate students [1].

The principal investigator relies on postdoctoral and graduate student workers motivated by the hope of achieving an independent research career and making important contributions, rather than monetary incentives. Students and postdoctoral fellows depend on the laboratory for education, career development, and income [1].

Biomedical research laboratories are classified into wet and dry laboratories (**Table 5-1**), with wet laboratories being the most extensively utilized space by individual investigators. Wet laboratories are defined as laboratories where chemicals, drugs, or other material or biological matter are tested and analyzed requiring water, direct ventilation, and specialized piped utilities [2].

Wet laboratory space types do not include biohazards in levels BL-2, BL-3, and BL-4 as defined by the 2009 NIH/CDC guideline "Biosafety in Microbiological and Biomedical Laboratories" [3]. Here NIH and CDC denote National Institutes of Health and Center for Disease Control and Prevention, respectively. They are two separate agencies under the United States Department of Health and Human Services (HHS).

The wet laboratory space types are typically located within a building specifically designed to house them. On the other hand, the dry laboratory space is a laboratory space that is specific to work with dry stored materials, electronics, and/or large instruments with few piped services. At research institutions, dry biomedical research laboratories are frequently the shared facilities by multiple faculty investigators within a department or research program.

Table 5-1. Wet and dry biomedical research laboratories [2]

TYPE OF LABORATORY	DESCRIPTION
Wet Laboratory	A wet laboratory requires working with solutions utilizing benches, sinks, etc., with fully piped services.
Dry Laboratory	A dry laboratory involves working with electronics and large instruments with few piped services.

3. Distinct Features of Biomedical Research Laboratories

Biomedical laboratories are distinguished by support space that is required for each laboratory or group of laboratories. Support space (i) houses shared equipment such as centrifuges, refrigerators, and freezers; (ii) provides spaces for functions that need to be separated and enclosed for environmental or safety reasons, such as cold rooms, warm rooms, and containment laboratories; or (iii) provides spaces for specialized functions, such as tissue culture, chemical, or infectious research laboratories. Autoclaving is appropriately conducted in support space.

A well designed and strategically located support space is critical to a successful biomedical research laboratory. The support space should be designed using the same design principles that are used in the laboratory, i.e., the same planning module [2]. Another distinguishing feature of biomedical laboratories is their need for proximity to live research objects, such as vivariums (animal housing) and clinics for human subjects; however this is not always the case.

The desk space for the scientific staff is very important. All laboratory staff should be provided with 1.5 m (5'-0") of desk length within the laboratory module, physically separated from the laboratory bench. The desk space does not have to be in the laboratory, as long as it is reasonably close to it [2].

4. General Guidelines on Planning and Designing of Biomedical Research Laboratories

The 2008 National Institutes of Health (NIH) Design Requirements Manual for Biomedical Laboratories and Animal Research Facilities provides detailed guidelines for designing biomedical laboratories at NIH [2]. This manual can also serve as a general guidance for designing biomedical research laboratories for faculty researchers at institutions of higher education, including those of traditionally non-research-intensive.

4.1. NIH Research Laboratory Design Minimum Requirements

An NIH biomedical research laboratory at a minimum requires the following basic components [2]:

- A hand washing sink with an emergency eyewash
- A safety shower where fume hoods are located or corrosives are handled as required
- An Underwriter Laboratories (UL) rated flammable storage cabinet
- An "internally explosion safe" refrigerator if required by the program
- A corrosive storage cabinet

4.2. Typical Design of Biomedical Research Laboratories

As described in the 2008 NIH Design Requirements Manual for Biomedical Laboratories and Animal Research Facilities, the following are typical designing requirements for wet biomedical research laboratories [2].

Wet biomedical research laboratories house functions that include working with solutions or biological materials and utilize benches, sinks, chemical fume hoods, and/or biological safety cabinets. Generally, a wet research laboratory is fitted out with a full range of piped services, such as deionized or reverse osmosis water, laboratory cold and hot water, laboratory waste/vents, carbon dioxide, vacuum, compressed air, hand washing sinks, eyewash, safety showers, natural gas, telephone, local area network (LAN), lighting, and power. Any wet research laboratory where biological specimens are used shall require an area to store medical pathologic waste. Listed below are additional important notions about wet research laboratories.

- Sufficient kneehole space shall be provided in each laboratory module to accommodate in use medical pathologic waste boxes as well as other in use waste receptacles.
- Laboratories that use radioactive materials will require a lockable storage area for multiple radioactive waste containers including dry and liquid waste.
- Access to wet and dry ice is required for most biomedical research.
- Work areas and desk space require low bench space with kneeholes or adjustable, flexible desktop space.
- These areas may be used for a large number of computers requiring heating, ventilation, and air conditioning, supplemental cooling, electricity, emergency power, uninterruptible power, and/or telecommunications/LAN.

4.3. Research Laboratory Space Requirements

4.3.1. General Principles of Laboratory Space Requirements

In planning and designing biomedical research laboratories, space requirements are among the most important issues to address. A dedicated institutional committee is usually needed to assess laboratory space requirements. Outlined below are the general principles regarding laboratory space requirements [2].

- Biomedical research includes a variety of scientific disciplines that need to be accommodated in the laboratory space.

- Laboratory facilities should provide space to perform experiments, electronic monitoring and calibration, information processing and retrieval, specimen and equipment storage, and recording equipment.
- Laboratories should be adaptable for the rapidly changing biomedical environment and research space needs, and should be able to support emerging scientific disciplines.
- Laboratory facilities should also provide space for administrative activities and informal staff interaction.
- Administrative space should include offices for the laboratory chiefs and their secretarial and support staff.
- Areas should be provided to encourage interaction and philosophical exchange of ideas between scientists. Interaction areas may include refreshment or break areas, copy centers, meeting rooms, corridors, and terraces.

4.3.2. General Rules of Thumb for Laboratory Space Planning

The general rules of thumb to be used in planning biomedical research laboratories for budget purposes, or when a laboratory program has not been developed, are provided in the 2008 NIH Design Requirements Manual for Biomedical Laboratories and Animal Research Facilities (**Table 5-2**) [2]. The laboratory support space is based on 50% of the laboratory. The U.S. Department of Health and Human Services' utilization rate for laboratories is 200 to 460 net square feet per person. The total area per person in the NIH table below is 342 net square feet.

Table 5-2. General rules of thumb for space planning of biomedical research laboratories [2]

SPACE	AREA (SQUARE METER)	AREA (SQUARE FEET)
Laboratory Space	16.5	181.5
Laboratory Support Space	8.25	90.5
Research Staff Office	2.79	30.0
Ancillary Space	0.84	9.0
Laboratory Administration	2.88	31.0
Optimal Area per Researcher Assigned Bench Space	31.26	342.0

5. Recommendations on Faculty Researcher Laboratory and Start Fund Allocation

5.1. General Considerations

High quality biomedical research needs high quality research laboratories. Each principal investigator shall be provided with adequate and comfortable laboratory space, including laboratory work space, laboratory support space, storage space, research space, and administrative support space in order to create a safe and functional research environment. For non-research-intensive institutions, each research-active faculty investigator should have a dedicated laboratory of sufficient space equipped with routine laboratory research equipment items, which include, at a minimum, tissue culture facility, regular centrifuge, microscope, carbon dioxide incubator, water bath, analytical balance, pH meter, and refrigerator, among others. These basic equipment items are used on a daily basis for biomedical research and are usually (if not exclusively) purchased through faculty researcher start fund and/or institutional budget (see Section 5.3 below on start fund).

5.2. Research Laboratory Size

The required size of the laboratory should be determined by considering factors including the scale and productivity of the research projects, the number of research fellows/graduate students working in the laboratory, and the size of the grant funding. For a junior faculty beginning his/her independent research career, a laboratory of approximately 400-600 net square feet is typically required at both research universities and traditionally non-research-intensive institutions engaged in biomedical research.

5.3. Start Fund for Research Laboratory Setup

5.3.1. Research-Intensive Institutions

One of the major keys to becoming successful for a new biomedical faculty member is to have sufficient resources at the beginning to start and maintain a biomedical research laboratory and to ensure time to gather data in support of grant proposals and apply for extramural funding. A new independent biomedical investigator should operate his or her research laboratory in much the same way that a CEO of a small business operates [1, 4, 5]. In this context, several excellent publications are available for guiding junior researchers to effectively utilize the resources, especially the start fund, and manage stress to build competitive biomedical research laboratories [6-9].

Start fund is the monetary package that is provided to new faculty researchers by the institutions that they just joined in. This fund is intended to allow the new faculty member to purchase the basic instrumentation and equipment necessary for his/her research, materials, and pay for the salaries of post-doctoral researchers and, if in a department or institution where their stipends are not defrayed by teaching, graduate students.

According to a recent editorial commentary published in Physiology by the American Physiology Society, a typical start-up package at a medical school basic science department including equipment, salary, and additional personnel usually exceeds $1.0 million. Throw in some renovations and other less transparent expenses, and by the time the principal investigator receives his/her first grant (and some never do) society has invested $2,000,000 ... not to mention someone's entire adult life [10]. Similarly, based on a report by Burroughs Welcome Fund [7], recent institutional offers of start-up support (excluding salary) for Ph.D.s averaged $800,000 (range $500,000 to $1,400,000), with a median of $750,000. Start-up offers for physician-scientists averaged $800,000 (range $100,000 to $2,110,250), with a median of $710,000. This money can be used for equipment, consumables, personnel costs, and professional travel. In many cases, the funds are made available for the first three years and normally (but not in all cases) may be carried forward if not spent.

5.3.2. Traditionally Non-Research-Intensive Institutions

Although the start fund for new faculty at research universities can be anywhere from \$500,000 to \$1,000,000 or more, depending on the research field and the type of research the faculty member will perform, it is important to note that faculty members at non-research based universities receive a much reduced start-up package, or do not receive one at all [11]. At the traditionally non-research-intensive institutions, including teaching universities and osteopathic medical schools that are determined to undertake the biomedical research endeavor, for all the newly recruited biomedical researchers regardless of their career stage, a start fund is needed to purchase the very basic and routinely utilized laboratory equipment items as listed above in Section 5.1 as well as certain specific equipment required by the ongoing or proposed future research projects. This is essential not only for the junior faculty members, but also for established investigators because the laboratory equipment is usually (if not exclusively) purchased via institutional fund/start package, and as such is generally not transferable. For those traditionally non-research-intensive institutions deemed to develop competitive biomedical research, a start fund ranging from \$100,000 to \$200,000 per researcher is necessary in order to attract talented young researchers with promising prospects.

Conclusion and Perspectives

As part of institutional biomedical research enterprise either core research facilities or faculty researchers' independent research laboratories are indispensible for research success. It is crucial that the institutional leadership not only appreciates the essentiality of individual research laboratories to the institution's biomedical research enterprise, but also understands the importance of efficient planning and designing of biomedical research laboratories as well as allocation of resources to attract and retain strong researchers and maximize the research output. While there is no universal formula to follow, the 2008 NIH Design Requirements Manual for Biomedical Laboratories and Animal Research Facilities [2] is a useful guidance for laboratory planning and designing. Certain successful strategies developed by respected research institutions might be also cautiously adopted as a reference to guide the allocation of laboratory space and startup fund. Lastly, it should be mentioned that it takes tremendous amounts of efforts and institutional funds to build full-scale buildings dedicated to biomedical research. As such, the concept of modular laboratories has recently been emerging as a cost-efficient strategy for establishing biomedical research laboratories at institutions that lack sufficient funds for building conventional research buildings [12-15].

References

[1] Freeman R, Weinstein E, Marincola E, Rosenbaum J, Solomon F. Competition and careers in biosciences. *Science.* 2001; 294:2293-4.

[2] Office of Research Facilities/NIH. The 2008 National Institutes of Health (NIH) Design Requirements Manual for Biomedical Laboratories and Animal Research Facilities (DRM).

http://orf.od.nih.gov/PoliciesAndGuidelines/BiomedicalandAnimalResearchFacilitiesDesignPoliciesandGuidelines/DRMHTMLver/DesignRequirementsManualTableofContents.htm (accessed on December 5, 2012).
[3] NIH/CDC. Biosafety in Microbiological and Biomedical Laboratories (BMBL), 5th edition (revised), 2009. http://www.cdc.gov/biosafety/publications/bmbl5/index.htm (accessed on December 5, 2012).
[4] Vogel G. A Day in the life of a topflight lab. *Science*. 1999; 285:1531-2.
[5] Burroughs Wellcome Fund. Academic tenure-track offer letters for new assistant professors in the biomedical sciences. http://www.bwfund.org/page.php?mode=privateview&pageID=350 (accessed on December 5, 2012).
[6] Carmichael ST. New laboratory start-up in the 21st century. *Trends Neurosci*. 2002; 25:287-8.
[7] Burroughs Wellcome Fund and Howard Hughes Medical Institute. Making the Right Moves: A Practical Guide to Scientific Management for Postdocs and New Faculty, 2nd edition, 2006. http://www.hhmi.org/resources/labmanagement/downloads/moves2.pdf (accessed on December 5, 2012).
[8] Andrade FH. Starting a new lab: developing new techniques and hiring personnel. http://www.the-aps.org/mm/Careers/Mentor/Managing-a-Lab-and-Personnel/Early-Career-Professionals/Supervising-staff-and-other-employees-/Developing-Techniques-and-Hiring-Personnel (accessed on December 5, 2012).
[9] Green DR. Stress in biomedical research: six impossible things. *Mol Cell*. 2010; 40:176-8.
[10] Boron WF. Sustainability in biomedical research. *Physiology* (Bethesda). 2010; 25:200-2.
[11] Here's what it actually costs to run a university science lab. http://io9.com/5827381/heres-what-it-actually-costs-to-run-a-university-science-lab (accessed on December 5, 2012).
[12] Modular laboratory buildings. http://www.germfree.com/articles/modular-laboratory (accessed on January 7, 2013).
[13] Quality modular laboratories. http://www.cpmlabfab.com (accessed on January 7, 2013).
[14] Stand alone labs: modular laboratory solutions for today's world. http://www.scientificfacilities.com/index.cfm?page=StandAloneLab (accessed on January 7, 2013).
[15] Modular laboratories specialized for BSL-3 and chemical research. http://lab-pro.com (accessed on January 7, 2013).

Self-Assessment Questions

[1] Biomedical research laboratories are classified into wet and dry labs. Which of the following best describes a dry lab?

A. A lab requiring the use of dry ice
B. A lab that only houses an instrument
C. Any lab if the bench is dry
D. Any lab if the floor is dry

[2] Which of the following most appropriately describes a wet biomedical lab?

A. A lab that involves working with electronics with few piped services
B. A lab that involves working with large equipment
C. A laboratory that requires working with solutions utilizing benches, sinks, etc., with fully piped services
D. Any lab if the floor is wet

[3] The 2008 U.S. National Institutes of Health (NIH) Design Requirements Manual for Biomedical Laboratories and Animal Research Facilities provides detailed guidelines for designing biomedical laboratories at the NIH. This manual can also serve as a general guidance for designing biomedical research laboratories for faculty researchers at non-NIH institutions. According to this manual, which of the following is not a required basic component for an NIH biomedical research laboratory?

A. A -40°C freezer
B. A corrosive storage cabinet
C. A hand washing sink with an emergency eyewash
D. A safety shower where fume hoods are located or corrosives are handled as required
E. An Underwriter Laboratories (UL) rated flammable storage cabinet

[4] A 3rd year D.O. student doing summer research is instructed to have her samples processed in a BL-3 lab. However, she has no idea about what BL-3 is. You tell her to refer to which of the following documents for answers?

A. The 2008 U.S. National Institutes of Health (NIH) Design Requirements Manual for Biomedical Laboratories and Animal Research Facilities (DRM)
B. The ICH Good Clinical Practice guidance (1996)
C. The 1999 NIH/CDC guideline "Biosafety in Microbiological and Biomedical Laboratories"
D. The 8th Edition of the Guide for the Care and Use of Laboratory Animals, NRC, 2011
E. The U.S. FDA Code of Federal Regulations 21CFR58: Good Laboratory Practice for Preclinical Studies

[5] Which of the following most appropriately describes the principles of biomedical research lab space requirement?

A. Laboratories should be adaptable for the rapidly changing biomedical environment and research space needs, and should be able to support emerging scientific disciplines.
B. Laboratory facilities should not provide space for administrative activities and informal staff interaction because of safety concern.

C. Laboratory facilities should not provide space for specimen storage due to federal regulations.
D. No areas should be provided to encourage interaction and philosophical exchange of ideas between scientists due to the confidentiality of the research findings.

Unit III.
Supportive Offices and Committees for Strategic Biomedical Research

Chapter 6

Office of Contracts and Grants Administration

Abstract

Office of contracts and grants administration, also known as office for sponsored research or office of sponsored programs is an indispensable component of strategic biomedical research at any institution of higher education that is engaged in research supported by extramural funding. In essence, office of contracts and grants administration serves as an irreplaceable liaison between researchers and sponsors (grant funding agencies) to convey and ensure compliance with internal and external regulations, policies, and procedures. This chapter describes the duties of office of contracts and grants administration and its relationship to department of research administration. This chapter is also intended to introduce the most current (2012) National Institutes of Health (NIH) grant policy statement as well as some common terms on grant application and administration. The chapter ends with an overview of the NIH grants process, which has been adopted by many other federal and non-federal funding agencies in the United States as well as many other developed countries.

Chapter at a Glance

1. Overview

Office of contracts and grants administration, also known as office for sponsored research, serves as an irreplaceable liaison between researchers and grant funding agencies (sponsors) to convey and ensure compliance with pertinent policies and regulations. Having in house personnel experienced in grant administration and management is indispensable for the institution to obtain competitive extramural grant funding, especially the National Institutes of Health (NIH) grants. As such, the institutional office of contracts and grants administration needs to be established if there is no one on the campus of the affiliated university. So do such committees as Institutional Animal Care and Use Committee (IACUC) and Institutional Review Board (IRB) (see Chapters 7 and 8 for detailed discussion on IACUC and IRB), which are responsible for reviewing and approving animal use protocol and human subjects research, respectively. The existence of these supportive offices/committees is not only required by federal/NIH policies, their ready availability also greatly increases the efficiency of faculty research efforts, and as such is a crucial factor in determining whether the research can progress and the institution can prosper. At the early stage of establishing and developing a research institution, these supportive offices and committees may not need to be operative on a full-time basis considering the limited institutional budget as well as the relatively lower scale of research activities at the beginning. In this context, the office for contracts and grants administration may be operated on a part-time basis. Likewise, the animal use protocols and the human subjects research may be reviewed by a local institution that provides such services.

2. Duties of Office of Contracts and Grants Administration

The office of contracts and grants administration (frequently called grant office for simplicity) provides assistance to faculty and staff in obtaining and managing sponsored awards that support scholarly activities while assuring proper stewardship of those funds. The office is charged with review and approval of proposals sent to all sponsors. The office also works closely with faculty and business office staff to ensure financial compliance and proper accounting of externally sponsored funds. The financial responsibilities include financial reporting and closeout, effort reporting, and development of the facilities and administrative cost rate. In essence, office of contracts and grants administration serves as a liaison between scientists, administrators, and the sponsors to convey and ensure compliance with internal and external regulations, policies, and procedures. The duties generally include the following:

- Proposal review and submission
- Examination and interpretation of and adherence to federal, state, and private foundation rules and regulations to safeguard sponsored research
- Negotiation, execution, and daily administration of grant awards and contracts
- Assistance with budget and budget forecasts
- Account establishment and monitoring

- Response to and execution of post award requirements and activities
- Providing updated information on funding opportunities and also (mainly in major research universities) providing training on grant writing for junior research faculty

3. Office of Contracts and Grants Administration versus Department of Research Administration

In major research institutions, office of contracts and grants administration is often part of the department of research administration (**Table 6-1**), which also includes office of research policy and compliance and office of patents, startups, and technology transfer among others [1-5]. In addition, pre-award and post-award businesses are frequently handled by separate offices. It should be noted that there are extensive interactions between these offices, and between the department of research administration and IACUC or IRB (**Figure 6-1**). As such, it is essential to have mechanisms at the institutional level to ensure adequate communications among these offices and committees with regard to providing support and guidance to researchers. Another notion is that in traditionally non-research-intensive universities, the duties listed in **Table 6-1** are usually fulfilled by personnel in a single office of contracts and grants administration. As such, the director of the office of contracts and grants administration must be experienced in the various areas of research administration and have a compelling vision to integrate the available expertise and resources to maximize the output.

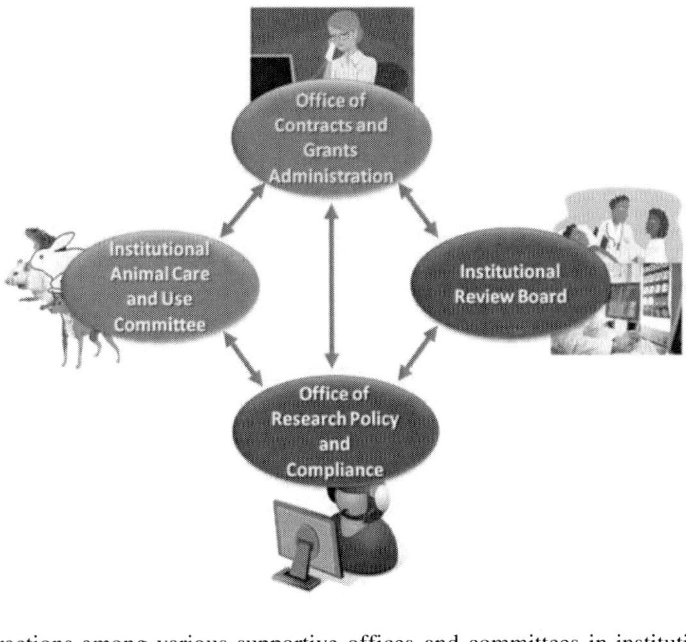

Figure 6-1. Interactions among various supportive offices and committees in institutional biomedical research. Note: office of contracts and grants administration is frequently called grant office for simplicity. At traditionally non-research-intensive institutions, grant office personnel may also be involved in activities pertinent to research policy and compliance, patents, and technology transfer among other responsibilities.

Table 6-1. Department of research administration and its responsibilities

Office	Responsibilities
Contracts and Grants Administration	To assist the campus research enterprise in applying for, receiving, and administering contracts and grants
Research Policy and Compliance	To interpret and develop research policy, coordinate issues that bridge compliance boundaries, and provide the research community with advice and assistance on matters related to research policy, ethics, and compliance
Extramural Fund Management	To manage and oversee all extramural research funds; in many institutions, this may be part of the office of contracts and grants administration.
Patents, Startups, and Technology Transfer	To protect, license, and transfer all campus inventions and intellectual property rights that have commercial and research value

4. National Institutes of Health (NIH) Grant Policy Statement

4.1. General Considerations

As discussed earlier in Chapter 2, the governmental support of biomedical research is predominantly by National Institutes of Health (NIH grants) though economic constrains are causing an NIH funding crisis [6-9]. As such, office of contracts and grants administration, at a minimum, should be familiar with the current NIH Grants Policy Statement and be able to provide assistance to faculty researchers in this regard.

The NIH Grants Policy Statement (NIHGPS) is intended to make available to NIH grantees, in a single document, the policy requirements that serve as the terms and conditions of NIH grant awards. This document also is designed to be useful to those interested in NIH grants by providing information about NIH: its organization, its staff, and its grants process. The 2012 version of NIHGPS is available online at the NIH website [1]. This version includes many links within the document as well as links to some web resources outside of this document. Users are strongly encouraged to use the on-line version of this document to benefit from these links.

4.2. Contents

The NIHGPS has three parts, which allows general information, application information, and other types of reference material to be separated from legally binding terms and conditions [10].

4.2.1. Part I: NIH Grants- General Information

Part I (chapters 1 and 2) contains a glossary defining commonly used terms and abbreviations used throughout the document; describes NIH and its relationship to other organizations within the Department of Health and Human Services (HHS); specifies grantee, NIH and other HHS staff responsibilities; and outlines the grant application and review processes.

4.2.2. Part II: Terms and Conditions of NIH Grant Awards

Part II (chapters 3-19) includes generally applicable terms and conditions (Part IIA). This part also specifies the terms and conditions that apply to particular types of grants, grantees, and activities that differ from, supplement, or elaborate on the standard terms and conditions (Part IIB). These requirements, in separate chapters, pertain to multiple program director/principal investigator (PD/PI) applications and awards; construction, modernization and major alteration and renovation grants; research training grants and fellowships; career development awards; modular applications and awards; conference grants, consortium agreements; grants to foreign and international organizations (and grants with substantial foreign components awarded to domestic organizations), grants to federal institutions and payments to federal employees; grants to for-profit organizations; and research patient care activities.

4.2.3. Part III: Points of Contact

Part III (chapter 20) lists pertinent offices with their contact information, including offices at NIH as well as other Health and Human Services (HHS) and Government offices.

4.3. Terms Pertaining to NIH Grant Application and Administration

NIH grants are the most competitive funding to obtain, and the mechanisms of NIH grant application and funding have been adopted widely by many government as well as private funding agencies. To help the readers, especially the junior faculty researchers and less experienced administrators, understand the mechanisms of NIH grant funding, **Table 6-2** lists some commonly encountered terms pertinent to NIH grant application and administration. The readers are referred to the comprehensive glossary and acronym list provided by the NIH Office of Extramural Research for more details [11].

Table 6-2. Common terms related to NIH grant application and administration [11]

TERM	DEFINITION
Grant	Financial assistance mechanism providing money, property, or both to an eligible entity to carry out an approved project or activity. A grant is used whenever the NIH Institute or Center anticipates no substantial programmatic involvement with the recipient during performance of the financially assisted activities.
Contract	An award instrument used to acquire from a non-federal party, by purchase, lease, or barter, property or services for the direct benefit or use of the Federal government. The same term may be used to describe a vendor relationship between a recipient and another party under a grant (to acquire routine goods and services); however, the recipient may use subaward to describe the contracts under a grant relationship.
Academic Research Enhancement Award (commonly known as AREA or R15 grant)	The AREA program supports small-scale research projects in the biomedical and behavioral sciences conducted by faculty and students at educational institutions that have not been major recipients of NIH research grant funds (with not more than $6 million per year of NIH support in total costs in each of four or more of the last seven years).

Table 6-2. (Continued)

TERM	DEFINITION
	The goals of R15 are to: Support meritorious research Expose students to research Strengthen the research environment of the institution Key features of R15: Project period is limited to 3 years. Direct costs are limited to $300,000 over the entire project period.
R01 Grant	The Research Project (R01) grant is an award made to support a discrete, specified, circumscribed project to be performed by the named investigator(s) in an area representing the investigator's specific interest and competencies, based on the mission of the NIH. It is the original and historically oldest grant mechanism used by NIH, and is one of the most competitive grant mechanisms. Applications for an R01 award are not limited in dollars but need to reflect the actual needs of the proposed project. Modular applications are most prevalent with modules of $25,000, up to the modular limit of $250,000. U.S. applicants requesting more than $250,000 in annual direct costs and all foreign applicants must complete and submit detailed budget requests using the Research & Related Budget component found in the application package.
Research Facilities Improvement Program (RFIP)	The National Center for Research Resources supports the Research Facilities Improvement Program (RFIP). RFIP grants are provided to public and nonprofit private biomedical institutions to expand, remodel, and renovate or alter existing research and animal facilities or construct new research and animal facilities. Improvements under this program must support basic and/or clinical biomedical and behavioral research, and they may also support research training.
Program Official (PO)	The NIH official responsible for the programmatic, scientific, and/or technical aspects of a grant.
Scientific Review Officer (SRO)	A Scientific Review Officer (SRO) is the NIH official who serves as the designated Federal official having legal responsibility for managing the peer review meeting, the procedures for evaluating the applications assigned to the Scientific Review Group and the determinations and management of conflicts of interest.
Center for Scientific Review (CSR)	The NIH component responsible for the receipt and referral of grant applications to the PHS, as well as the initial review for scientific merit of most applications submitted to the NIH.
eCommon	A secure meeting place on the web where research organizations and grantees electronically receive and transmit information about the administration of biomedical and behavioral research grants. Registration is required. At this site: Applicants access the status of their applications. Grantees access the status of their awards, submit reports, and make requests electronically.

TERM	DEFINITION
Grant.gov	Grants.gov (http://www.grants.gov) has been designated by the Office of Management and Budget as the single access point for all grant programs offered by 26 Federal grant-making agencies. It provides a single interface for agencies to announce their grant opportunities and for all applicants to find and apply for those opportunities.
Grantee	The organization or individual awarded a grant or cooperative agreement by NIH that is responsible and accountable for the use of the funds provided and for the performance of the grant-supported project or activity. The grantee is the entire legal entity even if a particular component is designated in the notice of award (NoA). The grantee is legally responsible and accountable to NIH for the performance and financial aspects of the grant-supported project or activity. Grantee is also known as awardee or recipient.
Signing Official (SO)	A Signing Official (SO) has institutional authority to legally bind the institution in grants administration matters. The individual fulfilling this role may have any number of titles in the grantee organization. The label, "Signing Official," is used in conjunction with the NIH eRA Commons. The SO can register the institution, and create and modify the institutional profile and user accounts. The SO also can view all grants within the institution, including status and award information. An SO can create additional SO accounts as well as accounts with any other role or combination of roles. For most institutions, the Signing Official (SO) is located in its Office of Sponsored Research or equivalent.
Authorized Organization Representative (AOR)	The individual, named by the applicant organization, who is authorized to act for the applicant and to assume the obligations imposed by the Federal laws, regulations, requirements, and conditions that apply to grant applications or grant awards. This official is equivalent to the Signing Official in the eRA Commons, i.e., holds the SO Role. Responsibilities of AOR include (i) submitting the grant on behalf of the company, organization, institution, or Government, and (ii) signing grant applications and the required certifications and/or assurances necessary to fulfill the requirements of the grant application process.
Program Director/Principal Investigator (PD/PI)	The individual(s) designated by the applicant organization to have the appropriate level of authority and responsibility to direct the project or program to be supported by the award. The applicant organization may designate multiple individuals as program directors/principal investigators (PD/PIs) who share the authority and responsibility for leading and directing the project, intellectually and logistically. When multiple PD/PIs are named, each is responsible and accountable to the applicant organization, or as appropriate, to a collaborating organization for the proper conduct of the project or program including the submission of all required reports. The presence of more than one PD/PI on an application or award diminishes neither the responsibility nor the accountability of any individual PD/PI.
Multiple Program Director/principal Investigator (Multiple PD/PI)	Multiple PD/PI awards are an opportunity for multidisciplinary efforts and collaboration through a team of scientists under a single grant award. All PD/PIs share equally the authority and responsibility for leading and directing the project, intellectually and logistically. Each PD/PI is responsible and accountable to the applicant organization, or as appropriate to a collaborating organization, for the proper conduct of the project or program, including the submission of all required reports. The presence of more than one PD/PI on an application or award diminishes neither the responsibility nor the accountability of any individual PD/PI.

Table 6-2. (Continued)

TERM	DEFINITION
Contact PD/PI	When multiple PD/PIs are designated, NIH requires that the applicant organization identify one of the PD/PIs as the Contact PD/PI to serve as a primary point of contact. Serving as Contact PD/PI confers no special authorities or responsibilities within the project team. The Contact PD/PI must meet all eligibility requirements for PD/PI status. However, as with the single PD/PI model, if the Contact PD/PI is not an employee, the applicant organization must have a formal written agreement with the Contact PD/PI that specifies an official relationship between the parties.
New Investigator	A PD/PI who has not previously competed successfully as a PD/PI for a substantial independent research award is considered a New Investigator.
Senior/Key Personnel	The PD/PI and other individuals who contribute to the scientific development or execution of a project in a substantive, measurable way, whether or not they receive salaries or compensation under the grant. Typically, these individuals have doctoral or other professional degrees, although individuals at the masters or baccalaureate level may be considered senior/key personnel if their involvement meets this definition. Consultants and those with a postdoctoral role also may be considered senior/key personnel if they meet this definition. "Zero percent" effort or "as needed" is not an acceptable level of involvement for senior/key personnel.
Co-Investigator	An individual involved with the PD/PI in the scientific development or execution of a project. The co-investigator (collaborator) may be employed by, or be affiliated with, the applicant/grantee organization or another organization participating in the project under a consortium agreement. A co-investigator typically devotes a specified percentage of time to the project and is considered senior/key personnel. The designation of a co-investigator, if applicable, does not affect the PD/PI's roles and responsibilities as specified in the NIH Grants Policy Statement (NIH GPS), nor is it a role implying multiple PD/PI.
Consultant	An individual who provides professional advice or services for a fee, but normally not as an employee of the engaging party. In unusual situations, an individual may be both a consultant and an employee of the same party, receiving compensation for some services as a consultant and for other work as a salaried employee.
Direct Costs	Costs that can be identified specifically with a particular sponsored project, an instructional activity, or any other institutional activity, or that can be directly assigned to such activities relatively easily with a high degree of accuracy. A related term is "indirect costs".
Indirect Costs [also known as Facilities and Administrative Costs (F&A)]	Costs that are incurred by a grantee for common or joint objectives and cannot be identified specifically with a particular project or program. These costs are also known as "indirect costs."
Total Project Costs	The total allowable costs (both direct costs and facilities and administrative costs) incurred by the grantee to carry out a grant-supported project or activity. Total project costs include costs charged to the NIH grant and costs borne by the grantee to satisfy a matching or cost-sharing requirement.
Dual Review System	Peer review process used by NIH. The first level of review provides a judgment of scientific merit. The second level of review (usually conducted by an ICD's advisory Council) assesses the quality of the first review, sets program priorities, and makes funding recommendations.

TERM	DEFINITION
Scientific Review Group (SRG) (also known as Study Section)	A peer review committee group of primarily nongovernment experts (peer reviewers), qualified by training or experience in particular scientific or technical fields, or as authorities knowledgeable in the various disciplines and fields related to the applications under review, to evaluate and give expert advice on the scientific and technical merit of the applications.
Streamlined Review	In the peer review process, applications not considered by the Scientific Review Group to be in the more meritorious half are "streamlined" and designated Not Discussed. Streamlined applications are not discussed at the review meeting and will not be assigned a numerical overall impact/priority score, but the applicants do receive the reviewers' critiques. Streamlined applications will receive criterion scores from the assigned reviewers in addition to the reviewers' critiques to help applicants assess whether or not they should submit a resubmission application.
Scored Review Criteria	For research grant applications, and most other types of applications, reviewers judge the overall impact to reflect their assessment of the likelihood for the project to exert a sustained, powerful influence on the research field(s) involved, taking into account five criteria, among other pertinent factors: Significance, Investigator(s), Innovation, Approach, and Environment. These scored review criteria may not be applicable for some types of applications. When these criteria are not applicable, the funding opportunity announcement (FOA) will include the specific review criteria. Reviewers will consider each of the five criteria in the determination of scientific and technical merit, and give a separate score for each. An application does not need to be strong in all categories to be judged likely to have a major scientific impact.
Summary Statement	A combination of the reviewers' written comments and the SRO's summary of the members' discussion during the study section meeting. It includes the recommendations of the study section, a recommended budget, and administrative notes of special considerations.
Impact/Priority Score	The impact/priority score is the rating which is assigned to an individual application by a Scientific review Group (SRG; also called study section), and designates the reviewers' assessment of the likelihood for the project to exert a sustained, powerful influence on the research field(s) involved, in consideration of established review criteria. The impact/priority score is one mechanism by which the SRG makes a recommendation to the funding component concerning the application's scientific and technical merit. Impact/priority scores may be numeric (10 – 90) or alphabetical (ND, for example; ND stands for not discussed).
Just-in-time (JIT)	NIH policy allows the submission of certain elements of a competing application to be deferred until later in the application process, after review when the application is under consideration for funding. This process is known as "Just-in-Time". Within the Status module of the eRA Commons, users will find a feature to submit Just-In-Time information when requested by the NIH. Through this module, institutions can electronically submit the information that is requested after the review, but before award.
Advisory Council/Board	National Advisory Council or Board, mandated by statute, providing the second level of review for grant applications for each Institute/Center awarding grants. The Councils/Boards are comprised of both scientific and lay representatives. Council/Board recommendations are based on scientific merit (as judged by the initial review groups) and the relevance of the proposed study to an institute's programs and priorities. With some exceptions, grants cannot be awarded without recommendations for approval by a Council/Board.

Table 6-2. (Continued)

TERM	DEFINITION
Payline	A percentile-based funding cutoff point determined by balancing the projected number of applications coming to an NIH Institute with the amount of funds available. Set after the budget is determined, paylines are not mandatory, are not made for all activity codes, and may be adjusted during the year.
Notice of Award (NoA)	The official, legally binding document, signed (or the electronic equivalent of signature) by a Grants Management Officer that (i) notifies the recipient of the award of a grant; (ii) contains or references all the terms and conditions of the grant and Federal funding limits and obligations; and (iii) provides the documentary basis for recording the obligation of Federal funds in the NIH accounting system.
Progress Report	Periodic, usually annual, report submitted by the grantee and used by NIH to assess progress and, except for the final progress report of a project period, to determine whether to provide funding for the budget period subsequent to that covered by the report. This report may also be called the non-competing continuation progress report.
No-Cost Extension	An extension of time to a project period and/or budget period to complete the work of the grant under that period, without additional Federal funds or competition. The first no-cost extension is "automatic" requiring no justifications; however, any additional (second) extension requires that justifications be provided to NIH.
Grant Closeout	The process by which a Federal awarding agency determines that all applicable administrative actions and all required work under an award have been completed by the grantee and the Federal awarding agency.

Note to **Table 6-2**: The terms are not listed alphabetically. Terms related to the general aspects of grants are listed first, followed by those related to grant administration at the NIH, principal investigators/institutions, grant review, notice of award, project progress report, and grant closeout. This order is largely in line with the overall flow of the NIH grants process (**Figure 6-2**).

5. NIH Grants Process Overview

The institutional office of contracts and grants administration should also provide or organize training workshops on opportunities of grant funding, grant types, and grant writing and application to assist faculty and staff in their pursuit and administration of external funding. These activities are especially important for junior faculty researchers who usually do not have enough experience in grant writing and application. Any successful project requires planning, development, implementation, and follow-through. Obtaining NIH funding for your research idea is no exception. **Figure 6-2** illustrates the overall grants process for obtaining NIH funding [12]. It provides an overview of the steps required for an application to proceed from application planning and submission through award and close out. This general scheme of NIH grants process is also adopted by many other federal and non-federal funding agencies in the United States as well as in other countries. Paying particular attention to special guidance from NIH experts can help maximize your understanding of the grants process and help you submit a successful grant application. The major types of NIH grants are listed in **Table 6-3** [13].

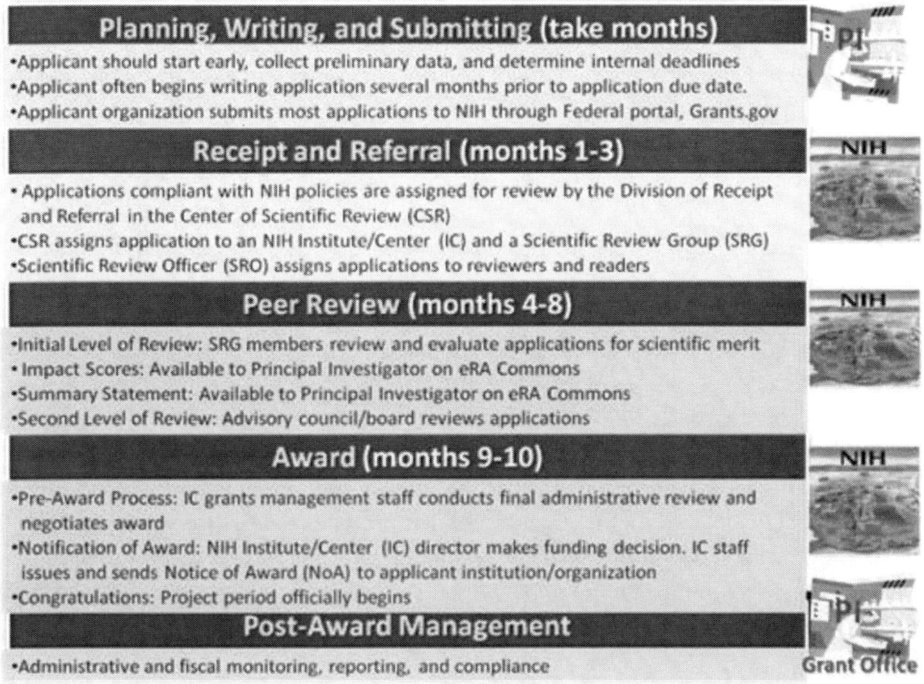

Figure 6-2. Schematic illustration of the overall NIH grants process [12]. It usually takes about one year or even longer from writing and submitting the proposal to NIH to finally (if successful) receiving the funding.

Table 6-3. Some of the major types of NIH grant programs [13]

TYPE	DESCRIPTION
Research Grants (R Series)	R01: NIH Research Project Grant Program R03: NIH Small Grant Program R13: NIH Support for Conferences and Scientific Meetings R15: NIH Academic Research Enhancement Award (AREA) R21: NIH Exploratory/Developmental Research Grant Award R34: NIH Clinical Trial Planning Grant Program R41/R42: Small Business Technology Transfer (STTR) R43/R44: Small Business Innovative Research (SBIR) R56: NIH High Priority, Short-Term Project Award
Program Project/Center Grants (P Series)	P01: Research Program Project Grant P20: Exploratory Grants P30: Center Core Grants P50: Specialized Center
Resource Grants	R24: Resource-Related Research Projects R25: Education Projects X01: Resource Access Program
Career Development Awards (K Series)	See detail at http://grants.nih.gov/training/careerdevelopmentawards.htm (accessed on December 10, 2012)
Research Training and Fellowships (T and F Series)	See detail at http://grants.nih.gov/training/nrsa.htm (accessed on December 10, 2012)

Conclusion and Perspectives

Office of contracts and grants administration, also known as office of sponsored programs (OSP), provides support to faculty researchers at institutions of higher education and has become an essential component of strategic biomedical research enterprise at major research universities as well as traditionally teaching institutions that are engaged in research. OSP identifies funding opportunities and partnerships, facilitates the submission of sponsored research project proposals on behalf of the institution/principal investigators, negotiates and executes agreements, provides advice, and guidance regarding applicable rules and regulations of grant funding. The functionality of the office of contracts and grants administration greatly influences the activities and success of grant applications from faculty researcher. While OSP is an irreplaceable entity of biomedical research enterprise at all research universities, its essentiality to institutional biomedical research has not received enough attention by the leadership at many traditionally non-research-intensive institutions. It is critical for faculty researchers to work closely with the top administrators to convince the leadership that establishment of a functional OSP that provides the best service possible to the institutional faculty and staff in their pursuit and administration of external funding is one of the first steps in developing competitive biomedical research to improve the institution's prestige and value.

References

[1] Johns Hopkins University. Research administration. http://jhuresearch.jhu.edu/index.htm (accessed on December 10, 2012).
[2] Harvard Medical School. Research administration. http://hms.harvard.edu/content/research-administration (accessed on December 10, 2012).
[3] University of Pennsylvania. Office of Research Services. http://www.upenn.edu/researchservices (accessed on December 10, 2012).
[4] Yale University. Office of Research Administration. http://researchadministration.yale.edu (accessed on December 10, 2012).
[5] Duke University School of Medicine. Office of Research Administration. http://research.som.duke.edu (accessed on December 10, 2012).
[6] Raven K. NIH aims to facilitate extramural research through new grants. *Nat Med.* 2012; 18:1309.
[7] Hromas R, Abkowitz JL, Keating A. Facing the NIH funding crisis: how professional societies can help. *JAMA.* 2012 Nov 26:1-2. doi: 10.1001/jama.2012.45067.
[8] Kaiser J. Biomedical research. Panel votes to end prevention fund, cut economics studies, freeze NIH. *Science.* 2012; 337:399.
[9] Collins FS. NIH basics. *Science.* 2012; 337:503.
[10] NIH. NIH grants policy statement (10/1/2012). http://grants.nih.gov/grants/policy/nihgps_2012/nihgps_2012.pdf (accessed on December 10, 2012).
[11] Office of Extramural Research, NIH. Glossary and acronym list. http://grants.nih.gov/grants/glossary.htm (accessed on December 10, 2012).

[12] Office of Extramural Research of the NIH. Grants process overview. http://grants.nih.gov/grants/grants_process.htm (accessed on December 10, 2012).

[13] Office of Extramural Research of the NIH. Types of grant programs. http://grants.nih.gov/grants/funding/funding_program.htm (accessed on December 10, 2012).

Self-Assessment Questions

[1] Office of contracts and grants administration, also known as office for sponsored programs (OSP) is an indispensable component of strategic biomedical research at any institution of higher education that is engaged in research supported by extramural funding. Which of the following is not among the duties of an OSP?

A. Examination and interpretation of and adherence to federal, state, and private foundation rules and regulations to safeguard sponsored research
B. Negotiation, execution, and daily administration of grant awards and contracts
C. Proposal writing, especially for the section of specific aims
D. Providing updated information on funding opportunities and also providing training on grant writing for junior research faculty
E. Response to and execution of post award requirements and activities

[2] The U.S. National Institutes of Health (NIH) supports various grant mechanisms, including R15. Which of the following is not among the primary goals of the R15 mechanism?

A. Expose students to research
B. Strengthen the research environment of the institution
C. Support meritorious research
D. Support student research at major research universities

[3] With regard to the U.S. National Institutes of Health (NIH) grant applications, which of the following best describes the responsibilities of a principal investigator (PI)?

A. A PI is a person who has institutional authority to legally bind the institution in grants administration matters.
B. A PI is an individual designated by the applicant organization to have the appropriate level of authority and responsibility to direct the project or program to be supported by the award.
C. A PI is an individual, named by the applicant organization, who is authorized to act for the applicant and to assume the obligations imposed by the federal laws, regulations, requirements, and conditions that apply to grant applications or grant awards.
D. A PI is an NIH official responsible for the programmatic, scientific, and/or technical aspects of a grant.

E. A PI is an NIH official who serves as the designated federal official having legal responsibility for managing the peer review meeting, the procedures for evaluating the applications assigned to the Scientific Review Group and the determinations and management of conflicts of interest.

[4] How long it generally takes, at a minimum, from writing a U.S. National Institutes of Health (NIH) grant proposal to having it finally funded?

A. About 1-2 weeks
B. About 1-2 months
C. About 3-4 months
D. About 9-12 months

[5] Professor G.T., a physician scientist, is informed that nowadays the U.S. National Institutes of Health (NIH) grant applications need to be submitted electronically via Grants.gov. Since he used to submit his NIH grant applications by express mails, Professor G.T has no idea about Grant.gov, and he asks a person working at the grant office for help. The grant office person explains to Professor G.T. that Grants.gov is:

A. A peer review process, where applications not considered by the Scientific Review Group to be in the more meritorious half are "streamlined" and designated Not Discussed.
B. A registry and results database of publicly and privately supported clinical studies of human participants conducted around the world.
C. A secure meeting place on the Web where research organizations and grantees electronically receive and transmit information about the administration of biomedical and behavioral research grants.
D. The single access point for all grant programs offered by 26 U.S. Federal grant-making agencies. It provides a single interface for agencies to announce their grant opportunities and for all applicants to find and apply for those opportunities.

Chapter 7

Institutional Animal Care and Use Committee

Abstract

The concept of Institutional Animal Care and Use Committee (IACUC) to review and ensure animal welfare is now common practice in the animal research community in the United States and many other countries. The goal of each IACUC is to ensure the humane care and use of animals used in research and compliance with pertinent guidelines and regulations, while maintaining flexibility to best meet the unique needs of the institution. This chapter first provides a historic overview of the development of IACUC and then discusses its responsibilities and constitution. The chapter also covers animal welfare assurance as well as the United States government principles for the utilization and care of vertebrate animals used in testing, research, and training.

Chapter at a Glance

6. Conclusion and Perspectives
7. References
8. Self-Assessment Questions

1. Overview

Institutional Animal Care and Use Committee (IACUC) is of central importance to the application of laws to animal research in the United States. Most research involving laboratory animals is funded by the United States National Institutes of Health (NIH) or other federal agencies. The NIH Office of Laboratory Animal Welfare (OLAW) has been directed by law to develop policies that describe the role of IACUCs [1]. Every institution in the United States that uses animals for federally funded laboratory research must have an IACUC. Each local IACUC reviews research protocols and conducts evaluations of the institution's animal care and use, which includes the results of inspections of facilities that are required by law. Hence, IACUC functions to oversee and routinely evaluate the animal care and use program at an institution. Here the "animal care and use program" refers to all activities conducted by and at an institution that have a direct impact on the well-being of animals, including animal and veterinary care, policies and procedures, personnel and program management and oversight, occupational health and safety, IACUC functions, and animal facility design and management [2] (also see Chapter 4).

2. Historic Development of IACUC

2.1. Animal Care Policies before 1950s

Prior to the middle of the 20th century the responsibility for animals used in research in the United States was placed directly in the hands of the researchers, and the quality of animal care and animal welfare varied tremendously among research institutions. Even within the same school or institution, research laboratories had inconsistent animal care policies and standards of care [3].

2.2. Formation of Animal Care Panel and Publication of the Guide

In 1961, a group of veterinarians working for research institutions in the Chicago area formed the Animal Care Panel (ACP) [3]. The ACP appointed a committee charged with establishing animal care and use guidelines for research facilities. Their efforts led to the publication of the first edition (1963) of the Guide for the Care and Use of Laboratory Animals (commonly known as the Guide). Subsequent editions of this publication were supported by the NIH and developed under the auspices of the Institute of Laboratory Animal Resources (ILAR), which was subsequently renamed the Institute for Laboratory Animal Research. The National Academies Press, under the auspices of the National Research Council, published the most recent (eighth) edition in 2011 [2]. This single document serves

4. IACUC Constitution

4.1. General Principles

The general principles of IACUC constitution include the following [2, 5, 6]:

- According to the NIH policy on IACUC, the IACUC must consist of at least 5 members who are appointed by the institution's chief executive officer (CEO). If the CEO delegates appointment authority, the delegation must be specific and in writing.
- The appointed members must be qualified through experience and expertise to provide oversight for the institution's animal programs, facilities, and procedures.
- At a minimum the IACUC must include a veterinarian, a practicing scientist experienced in animal research, a person whose primary concerns are in a nonscientific area, and a person who is unaffiliated with the institution except as a member of the IACUC (sometimes referred to as a public member).
- An individual who qualifies to fill more than one of the specified categories may be appointed to do so, but the committee must still consist of at least five members.
- The unaffiliated member should have no discernable ties or ongoing affiliation with the institution, and may not be a member of the immediate family of a person who is affiliated with the institution. Immediate family members include parent, spouse, child, and sibling.
- Appointment of an individual who is unambiguously unaffiliated is the best way to fulfill the letter and spirit of this provision. PHS Policy incorporates the *Guide* which specifies that public members should not be laboratory animal users.
- The veterinarian on the IACUC must have direct or delegated program authority and responsibility for animal-related activities and therefore is always considered to be affiliated with the institution.
- If an appointed member who fulfills one of the required specified positions (i.e., scientist, nonscientist, veterinarian, or unaffiliated) leaves the committee so that that position is no longer filled, the IACUC is not properly constituted and may not conduct official business until a member who fulfills the required position is appointed by the CEO (or designee).

4.2. The Membership Requirements

The IACUC membership requirements are summarized in **Box 7-1** [2].

Box 7-1. IACUC membership requirements

According to the Guide for the Care and Use of laboratory Animals (8th edition, 2011), the IACUC membership includes the following [2]:

- a Doctor of Veterinary Medicine either certified (e.g., by ACLAM, ECLAM, JCLAM, KCLAM) or with training and experience in laboratory animal science and medicine or in the use of the species at the institution

- at least one practicing scientist experienced in research involving animals
- at least one member from a nonscientific background, drawn from inside or outside the institution
- at least one public member to represent general community interests in the proper care and use of animals

Note: ACLAM, ECLAM, JCLAM, and KCLAM stand for the American College of Laboratory Animal Medicine, the European College of Laboratory Animal Medicine, the Japanese College of Laboratory Animal Medicine, and the Korean College of Laboratory Animal Medicine, respectively.

5. Animal Welfare Assurance

5.1. Definition

Animal welfare assurance refers to the documentation from an institution assuring institutional compliance with the Public Health Service (PHS) Policy on Humane Care and Use of Laboratory Animals (revised August 2002) [5].

5.2. Principles

- No activity involving animals may be conducted or supported by the PHS until the institution conducting the activity has provided a written assurance acceptable to the PHS, setting forth compliance with the Public Health Service Policy on Humane Care and Use of Laboratory Animals (revised August, 2002) [5].
- Assurances shall be submitted to the Office of Laboratory Animal Welfare (OLAW), Office of the Director, National Institutes of Health. The assurance shall be signed by the institutional official. OLAW will provide the institution with necessary instructions and an example of an acceptable assurance.
- All assurances submitted to the PHS in accordance with the Public Health Service Policy on Humane Care and Use of Laboratory Animals (revised August, 2002) [5] will be evaluated by OLAW to determine the adequacy of the institution's proposed program for the care and use of animals in PHS-conducted or supported activities. On the basis of this evaluation OLAW may approve or disapprove the assurance, or negotiate an approvable assurance with the institution.
- Approval of an assurance will be for a specified period of time (no longer than five years) after which time the institution must submit a new assurance to OLAW.
- OLAW may limit the period during which any particular approved assurance shall remain effective or otherwise condition, restrict, or withdraw approval.
- Without an applicable PHS-approved assurance no PHS-conducted or supported activity involving animals at the institution will be permitted to continue.
- The assurance needs to provide detailed description on the following three areas (detailed in **Table 7-1**):

(1) Institutional program for animal care and use
(2) Institutional status
(3) IACUC

Table 7-1. Animal welfare assurance describing three areas [7]

AREA	DESCRIPTION AND REQUIREMENTS
Institutional Program for Animal Care and Use	The assurance shall fully describe the institution's program for the care and use of animals in PHS-conducted or supported activities. The PHS requires institutions to use the Guide for the Care and Use of Laboratory Animals *(8th edition, 2011)* [2] as a basis for developing and implementing an institutional program for activities involving animals. The program description must include the following: • a list of every branch and major component of the institution, as well as a list of every branch and major component of any other institution, which is to be included under the assurance • the lines of authority and responsibility for administering the program and ensuring compliance with the Public Health Service Policy on Humane Care and Use of Laboratory Animals (revised August 2002) • the qualifications, authority, and responsibility of the veterinarian(s) who will participate in the program and the percent of time each will contribute to the program • the membership list of the Institutional Animal Care and Use Committee(s) (IACUC) established in accordance with the requirements set forth in the Public Health Service Policy on Humane Care and Use of Laboratory Animals • the procedures which the IACUC will follow to fulfill the requirements set forth in the Public Health Service Policy on Humane Care and Use of Laboratory Animals • the health program for personnel who work in laboratory animal facilities or have frequent contact with animals • a synopsis of training or instruction in the humane practice of animal care and use, as well as training or instruction in research or testing methods that minimize the number of animals required to obtain valid results and minimize animal distress, offered to scientists, animal technicians, and other personnel involved in animal care, treatment, or use • the gross square footage of each animal facility (including satellite facilities), the species housed therein and the average daily inventory, by species, of animals in each facility • any other pertinent information requested by OLAW
Institutional Status	Each institution must assure that its program and facilities are in one of the following categories (also see Section 4.1 of Chapter 4): Category 1 - Accredited by the AAALAC. All of the institution's programs and facilities (including satellite facilities) for activities involving animals have also been evaluated by the IACUC and will be reevaluated by the IACUC at least once every six months, and reports prepared in accordance with the Policy. Category 2 - Evaluated by the Institution. All of the institution's programs and facilities (including satellite facilities) for activities involving animals have been evaluated by the IACUC and will be reevaluated by the IACUC at least once every six months, and reports prepared in accordance with the Public Health Service Policy on Humane Care and Use of Laboratory Animals. The most recent semi-annual report of the IACUC evaluation shall be submitted to OLAW with the assurance.

Table 7-1. (Continued)

AREA	DESCRIPTION AND REQUIREMENTS
IACUC	The assurance must include the names, position titles, and credentials of the IACUC chairperson and the members. The committee shall consist of not less than five members, and must meet the membership requirement listed in **Box 7-1**.

Conclusion and Perspectives

As noted in Chapter 4, animal research has greatly benefited the public health of the human population. The use of animals in biomedical research is a privilege and must be conducted responsibly. As opposed to human subjects research (Chapter 8), where the human subjects volunteer for the clinical studies and sign an informed consent, laboratory animals obviously do not voluntarily participate in the research that usually causes pain and distress.

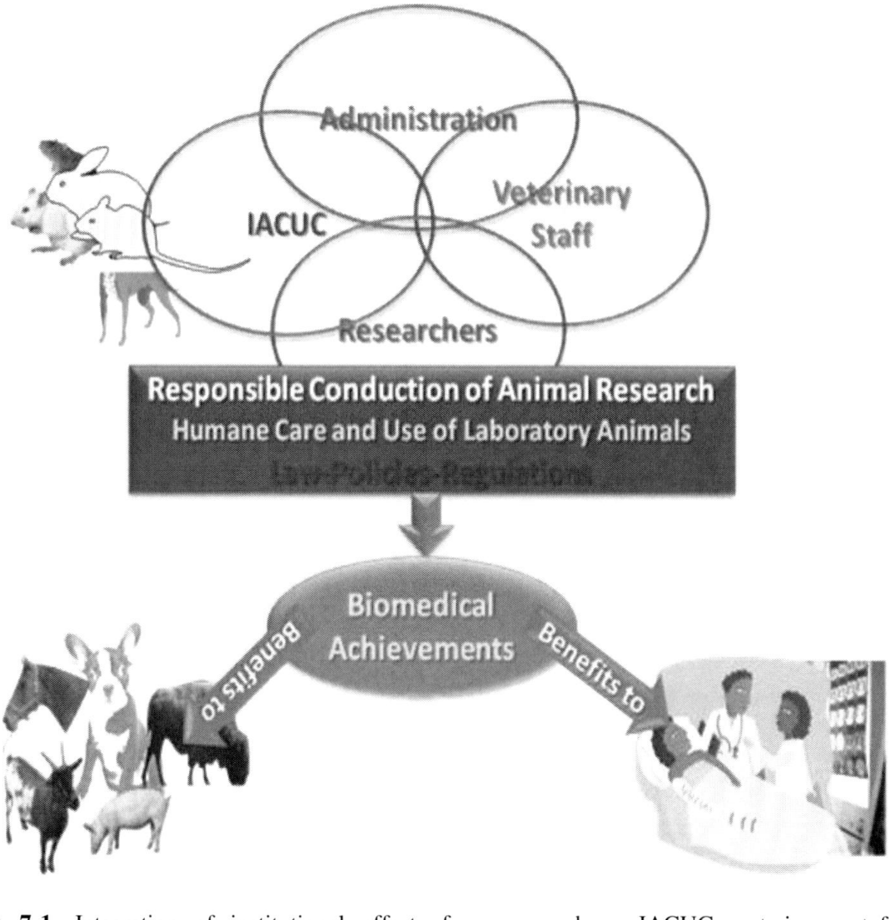

Figure 7-1. Integration of institutional efforts from researchers, IACUC, veterinary staff and administrators for responsible conduction of animal research to benefit both humans and non-human creatures. IACUC, Institutional Animal Care and Use Committee.

Although IACUC plays an important role in the humane care and use of laboratory animals in compliance with the United States federal law (Animal Welfare Act), the positive and collegial interactions and collaborations among biomedical researchers, veterinary staff, IACUC, and senior administrators are essential for the responsible conduction of animal research [8] (**Figure 7-1**). As animal research is an essential part of global biomedical research enterprises, developing countries need to adhere to the widely recognized ethical procedures and policies in conducting animal research [9-12]. In this context, the United States government principles for utilization and care of vertebrate animals used in testing, research, and training [13, 14] (**Box 7-2**) along with the 2011 Guide for the Care and Use of Laboratory Animals [2] provide a useful framework for developing institutional policies on humane care and use of laboratory animals and to meet the emerging challenges including issues on invertebrate research [15] as well as non-human primate research [16, 17] throughout the world. Nonhuman primates, in particular, are important for translational research because of their close physiological similarities to humans.

Box 7-2. The United States government principles for the utilization and care of vertebrate animals used in testing, research, and training [13, 14]

The development of knowledge necessary for the improvement of the health and well-being of humans as well as other animals requires in vivo experimentation with a wide variety of animal species. Whenever United States government agencies develop requirements for testing, research, or training procedures involving the use of vertebrate animals, the following principles shall be considered; and whenever these agencies actually perform or sponsor such procedures, the responsible institutional official shall ensure that these principles are adhered to:

- Principle 1: The transportation, care, and use of animals should be in accordance with the Animal Welfare Act (7 U.S.C. 2131 et. seq.) and other applicable federal laws, guidelines, and policies.
- Principle 2: Procedures involving animals should be designed and performed with due consideration of their relevance to human or animal health, the advancement of knowledge, or the good of society.
- Principle 3: The animals selected for a procedure should be of an appropriate species and quality and the minimum number required to obtain valid results. Methods such as mathematical models, computer simulation, and in vitro biological systems should be considered.
- Principle 4: Proper use of animals, including the avoidance or minimization of discomfort, distress, and pain when consistent with sound scientific practices, is imperative. Unless the contrary is established, investigators should consider that procedures that cause pain or distress in human beings may cause pain or distress in other animals.
- Principle 5: Procedures with animals that may cause more than momentary or slight pain or distress should be performed with appropriate sedation, analgesia, or anesthesia. Surgical or other painful procedures should not be performed on unanesthetized animals paralyzed by chemical agents.

- Principle 6: Animals that would otherwise suffer severe or chronic pain or distress that cannot be relieved should be painlessly killed at the end of the procedure or, if appropriate, during the procedure.
- Principle 7: The living conditions of animals should be appropriate for their species and contribute to their health and comfort. Normally, the housing, feeding, and care of all animals used for biomedical purposes must be directed by a veterinarian or other scientist trained and experienced in the proper care, handling, and use of the species being maintained or studied. In any case, veterinary care shall be provided as indicated.
- Principle 8: Investigators and other personnel shall be appropriately qualified and experienced for conducting procedures on living animals. Adequate arrangements shall be made for their in-service training, including the proper and humane care and use of laboratory animals.
- Principle 9: Where exceptions are required in relation to the provisions of these principles, the decisions should not rest with the investigators directly concerned but should be made, with due regard to Principle 2, by an appropriate review group such as an institutional animal care and use committee (IACUC). Such exceptions should not be made solely for the purposes of teaching or demonstration.

References

[1] NIH. Office of Laboratory Animal Welfare. http://grants.nih.gov/grants/olaw/olaw.htm (accessed on December 11, 2012).

[2] National Research Council/National Academies Press. Guide for the Care and Use of Laboratory Animals, 8th ed (2011), Washington D.C. http://www.nap.edu/catalog.php?record_id=12910 (accessed on December 11, 2012).

[3] OLAW/NIH. Institutional Animal Care and Use Committee Guidebook, 2nd edition (2002). http://grants.nih.gov/grants/olaw/guidebook.pdf (accessed on December 11, 2012).

[4] Cardon AD, Bailey MR, Bennett BT. The Animal Welfare Act: from enactment to enforcement. *J Am Assoc Lab Anim Sci.* 2012; 51:301-5.

[5] OLAW/NIH. Public Health Service Policy on Humane Care and Use of Laboratory Animals (2002). http://grants.nih.gov/grants/olaw/references/phspolicylabanimals.pdf (accessed on December 11, 2012).

[6] OLAW/NIH. The Institutional Animal Care and Use Committee. http://grants.nih.gov/grants/olaw/tutorial/iacuc.htm (accessed on December 11, 2012).

[7] OLAW/NIH. Animal welfare assurance. http://grants.nih.gov/grants/olaw/references/phspol.htm#AnimalWelfareAssurance (accessed on December 11, 2012).

[8] Couto M. Laboratory guidelines for animal care. *Methods Mol Biol.* 2011; 770:579-99.

[9] Meng X, Hamer R, Meng Q, Wang P, Meng F, Li H, Feng J, Xue D, Zhou Y. Animal welfare development in China. *Science.* 2012; 338:1150-1.

[10] Wells DJ. Animal welfare and the 3Rs in European biomedical research. *Ann N Y Acad Sci.* 2011; 1245:14-6.

[11] Nyika A. Animal research ethics in Africa: an overview. *Acta Trop.* 2009; 112(Suppl 1):S48-52.
[12] Ramalli EL Jr, Ho W, Alves M, Rocha EM. Progress in animal experimentation ethics: a case study from a brazilian medical school and from the international medical literature. *Acta Cir Bras.* 2012; 27:659-63.
[13] OLAW/NIH. U.S. government principles for the utilization and care of vertebrate animals used in testing, research, and training. http://grants.nih.gov/grants/olaw/tutorial/relevant.htm (accessed on December 11, 2012).
[14] U.S. Office of Science and Technology Policy. Laboratory animal welfare; U.S. government principles for the utilization and care of vertebrate animals used in testing, research and training; notice. *Fed Regist.* 1985; 50:20864-5.
[15] Harvey-Clark C. IACUC Challenges in invertebrate research. *ILAR J.* 2011; 52:213-20.
[16] Miller-Spiegel C. Primate studies: fix welfare issues first. *Nature.* 2012; 484:167.
[17] Wadman M. Activists ground primate flights. *Nature.* 2012; 483:381-2.

Self-Assessment Questions

[1] The Laboratory Animal Welfare Act, the first version of what is now known as the Animal Welfare Act (AWA), passed by the U.S. Congress (Public Law 89-544) establishes legal standards for laboratory animal care and use for the first time in this country. When was AWA passed?

A. 1816
B. 1908
C. 1966
D. 2001

[2] Institutional Animal Care and Use Committee (IACUC) is of central importance to the application of laws to animal research in the United States. The minimum number of members for an IACUC is:

A. 3
B. 5
C. 7
D. 9
E. 11

[3] Animal Welfare Assurance refers to the documentation from an institution assuring institutional compliance with the U.S. Public Health Service (PHS) Policy on Humane Care and Use of Laboratory Animals. Animal Welfare Assurance should be submitted to which of the following offices/organizations?

A. Office of Global Regulatory Operations and Policy at the U.S. Food and Drug Administration
B. Office of Laboratory Animal Welfare at the U.S. National Institutes of Health

C. Office of the Inspector General at the U.S. Department of Agriculture
D. The Association for Assessment and Accreditation of Laboratory Animal Care

[4] Which of the following is not among the responsibilities of an Institutional Animal Care and Use Committee (IACUC)?

A. Inspects, at least semiannually, the institution's animal facilities, including satellite facilities
B. Makes sure that the animal facility is the Association for Assessment and Accreditation of Laboratory Animal Care (AAALAC)-accredited
C. Reviews and approves those components of PHS conducted or supported activities related to the care and use of animals
D. Reviews animal welfare concern
E. Reviews, at least semiannually, the institution's program for the humane care and use of animals

[5] The Institutional Animal Care and Use Committee (IACUC) membership requirements are documented in the Guide for the Care and Use of Laboratory Animals (8th edition, 2011). According to the Guide, which of the following is an absolute membership requirement?

A. A female scientist working in the field of animal science
B. A PhD scientist working in the field of animal science
C. A physician who has animal research background
D. A safety officer trained specifically for animal facilities
E. An individual who is unaffiliated with the institute except as the member of the IACUC

Chapter 8

Human Subjects Research and Institutional Review Board (IRB)

Abstract

The use of human subjects in research benefits society in many ways, from contributing to the development of new drugs and medical procedures to understanding how we think and behave. It also can and has imposed unacceptable risks on research subjects. To help ensure that the risks do not outweigh the benefits, human subjects research is carefully regulated by society. This chapter is intended to introduce the basic ethical principles and federal regulations of human subjects research with an emphasis on the role of Institutional Review Board (IRB) in the protection of the human subjects in biomedical research.

Chapter at a Glance

1. Historic Overview of the Protection of Human Subjects in Research

1.1. First International Guidelines Following World War II

Society protects the welfare of individuals in many ways, but it did not specifically address the issue of the welfare of research subjects until after World War II. Following the War, widespread concerns about atrocities committed during the War in the name of research led to the formulation of a code for human subjects research known as the Nuremberg Code (1947). Although not binding on researchers, the Nuremberg Code and the later Declaration of Helsinki (1964; latest revision and clarification, 2002) provided the first explicit international guidelines for the ethical treatment of human subjects in research [1].

1.2. Public Concerns in 1950's/1960's Leading to U.S. Government Regulation

The Nuremberg Code and Declaration of Helsinki did not put an end to unethical human subjects research. During the Cold War, U.S. researchers tested the effects of radiation on hospital patients, children, and soldiers without obtaining informed consent or permission to do so. Through the 1950's and 1960's, well after antibiotics effective for the treatment of syphilis were discovered, scores of African-American males in a long-term syphilis study (conducted by the U.S. Public Health Service in Tuskegee, Alabama) were not offered treatment with the new drugs so that researchers could continue to track the course of the disease [2]. These and other questionable practices raised serious public concern and led eventually to U.S. government regulation on human subjects research.

1.3. U.S. Federal Regulations and the Belmont Report in 1970's

To prevent aforementioned and similar abuses of human subjects in research from continuing, in 1974 the U.S. Congress required the Department of Health, Education and Welfare (HEW, currently Health and Human Services- HHS) to clarify its rules for the use of human subjects in research. With this mandate in hand, HEW codified its procedures under Title 45 of the Code of Federal Regulations, Part 46 (45CFR46) [1]. (At roughly the same time, the U.S. Food and Drug Administration codified its rules for human subjects research under 21CFR50 and 56.) These early regulations have been revised for multiple times since their first issue date (see Section 2.1 for more discussion). Congress also called in 1974 for the creation of a National Commission for the Protection of Human Subjects of Biomedical and Behavioral Research. During the four years it met, the Commission issued a number of reports on the protection of research subjects and recommended principles for judging the ethics of human subjects research. The most notable one is the "Belmont Report: Ethical Principles and Guidelines for the Protection of Human Subjects of Research, Report of the National Commission for the Protection of Human Subjects of Biomedical and Behavioral Research". This report is commonly known as the Belmont report for short, issued in 1978 and published in 1979 [3]. The report took its name from the Belmont Conference Center (once a part of the Smithsonian Institution, Maryland, USA) where the document was drafted in part. The Belmont Report outlines the basic ethical principles in research involving human subjects. The current U.S. system of protection for human subjects in research is heavily influenced by the Belmont Report (see Section 2.1 for more discussion).

1.4. Adoption of the Common Rule in 1990's

In 1991 most U.S. federal departments and agencies that conduct or support human subjects research adopted a common set of regulations for the protection of human subjects referred to as the "Common Rule" (45CFR46, Subpart A) [4]. Additional requirements on three sensitive research areas are also included in 45CFR46:

- Subpart A: Known as the Common Rule describing the basic set of protections for all human subjects of research, including description of Institutional Review Board (IRB).
- Subpart B: Additional Protections for Pregnant Women, Human Fetuses and Neonates Involved in Research.
- Subpart C: Additional Protections Pertaining to Biomedical and Behavioral Research Involving Prisoners as Subjects.
- Subpart D: Additional Protections for Children Involved as Subjects in Research.

Together, 45CFR46, Subparts A-D, provide a comprehensive articulation of society's expectations for the responsible use of human subjects in research in the United States. Authority for enforcing the HHS regulations for the protection of human subjects who participate in research conducted or supported by HHS now rests with the Office for Human Research Protections (OHRP) in the Office of Public Health and Science (OPHS) of the U.S. Department of Health and Human Services [1, 5].

2. Human Subjects Research and the Protection of Human Subjects

2.1. Basic Ethical Principles

As noted earlier, the current U.S. system of protection for human research subjects is heavily influenced by the Belmont Report, published in 1979 by the National Commission for the Protection of Human Subjects of Biomedical and Behavioral Research. The Belmont Report outlines the basic ethical principles in research involving human subjects (**Box 8-1**). In 1981, with this report as foundational background, HHS and the Food and Drug Administration (FDA) revised, and made as compatible as possible under their respective statutory authorities, their existing human subjects regulations (see Sections 2-3 and 2-4).

Box 8-1. Basic ethical principles outlined in the Belmont Report [1, 3]

(1) **Respect:** Respect for persons and their right to make decisions for and about themselves without undue influence or coercion from someone else (the researcher in most cases).

(2) **Beneficence:** Beneficence or the obligation to maximize benefits and reduce risks to the subject.

(3) **Justice:** Justice or the obligation to distribute benefits and risks equally without prejudice to particular individuals or groups, such as the mentally disadvantaged or members of a particular race or gender.

2.2. United States Federal Policy for the Protection of Human Subjects

The Federal Policy for the Protection of Human Subjects or the "Common Rule" was published in 1991 and codified in separate regulations by 15 federal departments and agencies as listed in **Box 8-2** [4]. Each agency includes in its chapter of the Code of Federal Regulations [CFR] section numbers and language that are identical to those of the HHS codification at 45CFR part 46, subpart A (also known as the Common Rule). For all participating departments and agencies the Common Rule outlines the basic provisions for IRBs, informed consent, and assurances of compliance. Human subjects research conducted or supported by each federal department/agency is governed by the regulations of that department/agency.

Box 8-2. Federal departments and agencies that have signed to the Common Rule [4]

The Common Rule was based on the HHS 45CFR part 46 subpart A (see Section 1.4 above) and includes identical language in the separate regulations of those departments and agencies.

Technical amendments were made to the Common Rule in 2005. One additional agency (the Central Intelligence Agency) is required to follow the Common Rule by executive order, and one additional department (the Department of Homeland Security) chose to follow all HHS subparts. Other departments and agencies have adopted one or more of the other HHS subparts and some have their own additional human subject protection regulations.

The list below displays the 15 U.S. federal agencies and departments that have signed onto the Common Rule and their CFR numbers.

Adoption of the Common Rules by the 15 U.S. Federal Agencies and Departments (excluding The Central Intelligence Agency, The Department of Homeland Security, and The Social Security Administration)

7CFR Part 1c	Department of Agriculture
10CFR Part 745	Department of Energy
14CFR Part 1230	National Aeronautics and Space Administration
15CFR Part 27	Department of Commerce
16CFR Part 1028	Consumer Product Safety Commission
22CFR Part 225	Agency for International Development (USAID)
24CFR Part 60	Department of Housing and Urban Development
28CFR Part 46	Department of Justice
32CFR Part 219	Department of Defense
34CFR Part 97	Department of Education
38CFR Part 16	Department of Veterans Affairs
40CFR Part 26	Environmental Protection Agency
45CFR Part 46	Department of Health and Human Services
45CFR Part 690	National Science Foundation
49CFR Part 11	Department of Transportation

2.3. The HHS Regulations

As introduced in Section 1.4, the U.S. Department of Health and Human Services (HHS) regulations, 45CFR part 46, include four subparts: (i) subpart A, also known as the Federal Policy or the "Common Rule"; (ii) subpart B, additional protections for pregnant women, human fetuses, and neonates (issued in 1975, revised in 2001); (iii) subpart C, additional protections for prisoners (issued in 1978); and (iv) subpart D, additional protections for children issued in 1983). Recently, a subpart E has been included in the HHS regulations. Subpart E, issued in 2009, requires registration of Institutional Review Boards (IRBs) which conduct review of human research studies conducted or supported by HHS.

The Office for Human Research Protections (OHRP) at the U.S. Department of Health and Human Services provides leadership in the protection of the rights, welfare, and well-being of human subjects involved in research conducted or supported by the U.S. Department of Health and Human Services. OHRP helps ensure this by providing clarification and guidance, developing educational programs and materials, maintaining regulatory oversight, and providing advice on ethical and regulatory issues in biomedical and social-behavioral research. Before engaging in HHS-conducted or -supported human subjects research that is not exempt under HHS regulations at 45CFR46.101(b), an institution must fulfill the following requirements[6]:

- hold or obtain an OHRP-approved Federalwide Assurance (FWA); and,
- certify to the HHS agency conducting or supporting the research that the research has been reviewed and approved by an IRB designated in the FWA and will be subject to continuing review by an IRB.

2.4. The FDA Regulations

In addition to the well-known HHS regulations 45 CFR part 46 described above, the United States Food and Drug Administration (FDA), an HHS agency that regulates clinical investigations of products under its jurisdiction, such as drugs, biological products, and medical devices, also has its own regulations on human subjects research and IRB. The FDA regulations are published as part of chapter 21 of the CFR, and FDA's human subject protection regulations are in parts 50, 56, 312, and 812 [7, 8].

- 21CFR part 50 deals with protection of human subjects
- 21CFR part 56 deals with IRB
- 21CRF part 312 deals with investigational new drug application
- 21CRF part 812 deals with investigational device exemptions

2.5. Summary

Human subjects research conducted in the United States must be designed, reviewed, approved, and implemented in accord with accepted ethical principles and the U.S. Department of Health and Human Services (45CFR46) and U.S. FDA (21CFR50 and 56) regulations for the protection of human subjects. Hence, the IRB is described in the context of both HHS (45CFR46) and FDA (21 CFR50 and 56) regulations. In this regard, many aspects of HHS and FDA regulations on IRB are virtually identical [8]. Section 3 below provides an overview of IRB, including its definition and functions, criteria for IRB approval, and IRB membership requirements. The Federalwide Assurance (FWA), a written assurance of compliance, is covered in Section 4.

3. Institutional Review Board

3.1. Defining IRB and Human Subjects

An Institutional Review Board (IRB), also known as an institutional review committee, independent ethics committee, or ethical review board, is a committee formally designated by an institution to review, to approve the initiation of, and to conduct periodic review of, biomedical research involving human subjects. The primary purpose of such review is to assure the protection of the rights and welfare of the human subjects. As noted above, in the United States, the Food and Drug Administration (FDA) and Department of Health and Human Services (specifically the Office for Human Research Protections) regulations have empowered IRBs to approve, require modifications in planned research prior to approval, or disapprove research (**Figure 8-1**). IRBs are responsible for critical oversight functions for research conducted on human subjects that are scientific, ethical, and regulatory. Here, the term "human subject" is defined by the FDA (21CFR50) and the HHS (45CFR46) regulations (**Table 8-1**) [8]. As shown, both define "human subject" similarly in meaning, but with difference in wording.

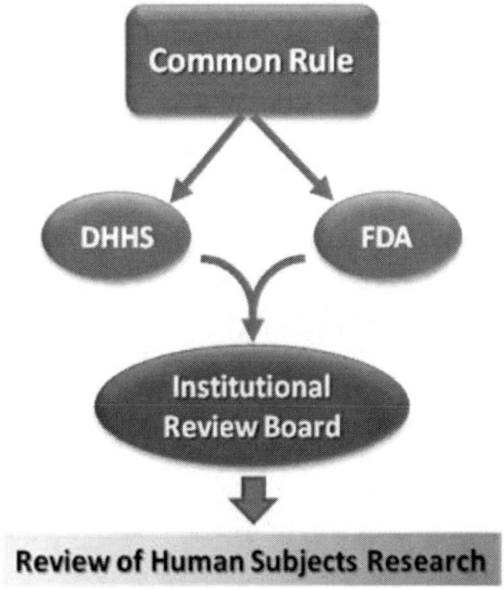

Figure 8-1. DHHS and FDA regulations empowering Institutional Review Board to review human subjects research. DHHS, Department of Health and Human Services.

Table 8-1. Definition of human subjects in the context of human subjects research [8]

FDA Definition	"Human subject" means an individual who is or becomes a participant in research, either as a recipient of the test article or as a control. A subject may be either a healthy individual or a patient. FDA has defined "clinical investigation" to be synonymous with "research". "Clinical investigation" means any experiment that involves a test article and one or more human subjects, and that either must meet the requirements for prior submission to the FDA or the results of which are intended to be later submitted to, or held for inspection by, the FDA as part of an application for a research or marketing permit.
HHS Definition	"Human subject" means a living individual about whom an investigator (whether professional or student) conducting research obtains: (i) data through intervention or interaction with the individual, or (ii) identifiable private information. HHS has defined "research" as a systematic investigation, including research development, testing, and evaluation, designed to develop or contribute to generalizable knowledge.

3.2. Functions of IRB

Both FDA and HHS regulations define the functions of IRB in a similar, but not identical, manner. Under FDA regulations (21CFR56), each IRB shall [9]:

- Follow written procedures (i) for conducting its initial and continuing review of research and for reporting its findings and actions to the investigator and the institution; (ii) for determining which projects require review more often than annually and which projects need verification from sources other than the investigator that no material changes have occurred since previous IRB review; (iii) for ensuring prompt reporting to the IRB of changes in research activity; and (iv) for ensuring that changes in approved research, during the period for which IRB approval has already been given, may not be initiated without IRB review and approval except where necessary to eliminate apparent immediate hazards to the human subjects.
- Follow written procedures for ensuring prompt reporting to the IRB, appropriate institutional officials, and the FDA of: (i) any unanticipated problems involving risks to human subjects or others; (ii) any instance of serious or continuing noncompliance with these regulations or the requirements or determinations of the IRB; or (iii) any suspension or termination of IRB approval.
- Except when an expedited review procedure is used, review proposed research at convened meetings at which a majority of the members of the IRB are present, including at least one member whose primary concerns are in nonscientific areas. In order for the research to be approved, it shall receive the approval of a majority of those members present at the meeting.

Under HHS (45CFR46) regulations, each IRB shall [10]:

- Follow written procedures (i) for conducting its initial and continuing review of research and for reporting its findings and actions to the investigator and the institution; (ii) for determining which projects require review more often than annually and which projects need verification from sources other than the investigators that no material changes have occurred since previous IRB review; and (iii) for ensuring prompt reporting to the IRB of proposed changes in a research activity, and for ensuring that such changes in approved research, during the period for which IRB approval has already been given, may not be initiated without IRB review and approval except when necessary to eliminate apparent immediate hazards to the subject.
- Follow written procedures for ensuring prompt reporting to the IRB, appropriate institutional officials, and the department or agency head of (i) any unanticipated problems involving risks to subjects or others or any serious or continuing noncompliance with this policy or the requirements or determinations of the IRB; and (ii) any suspension or termination of IRB approval.
- Except when an expedited review procedure is used, review proposed research at convened meetings at which a majority of the members of the IRB are present, including at least one member whose primary concerns are in nonscientific areas. In order for the research to be approved, it shall receive the approval of a majority of those members present at the meeting. Here the term "expedited review" refers to the review of research which involves no more than minimal risk to human subjects and in which the only involvement of human subjects is in one or more of the categories applicable to experdited review as listed in the Federal Register notice of Nov. 9, 1998 (http://www.hhs.gov/ohrp/policy/63fr60364.html; accessed on May 22, 2013).

3.3. Criteria for IRB Approval

The criteria for IRB approval under both the FDA and HHS regulations are virtually the same. In order to approve research covered by the FDA or HHS regulations, the IRB shall determine that all of the following requirements are satisfied [8-10]:

- Risks to subjects are minimized: (i) by using procedures which are consistent with sound research design and which do not unnecessarily expose subjects to risk, and (ii) whenever appropriate, by using procedures already being performed on the subjects for diagnostic or treatment purposes.
- Risks to subjects are reasonable in relation to anticipated benefits, if any, to subjects, and the importance of the knowledge that may reasonably be expected to result. In evaluating risks and benefits, the IRB should consider only those risks and benefits that may result from the research (as distinguished from risks and benefits of therapies subjects would receive even if not participating in the research). The IRB should not consider possible long-range effects of applying knowledge gained in the research (for example, the possible effects of the research on public policy) as among those research risks that fall within the purview of its responsibility.
- Selection of subjects is equitable. In making this assessment the IRB should take into account the purposes of the research and the setting in which the research will be conducted, and should be particularly cognizant of the special problems of research involving vulnerable populations, such as children, prisoners, pregnant women, mentally disabled persons, or economically or educationally disadvantaged persons.
- Informed consent will be sought from each prospective subject or the subject's legally authorized representative, in accordance with, and to the extent required by 45CFR46.116.
- Informed consent will be appropriately documented, in accordance with, and to the extent required by 45CFR46.117.
- When appropriate, the research plan makes adequate provision for monitoring the data collected to ensure the safety of subjects.
- When appropriate, there are adequate provisions to protect the privacy of subjects and to maintain the confidentiality of data.
- In addition, when some or all of the subjects are likely to be vulnerable to coercion or undue influence, such as children, prisoners, pregnant women, mentally disabled persons, or economically or educationally disadvantaged persons, additional safeguards have been included in the study to protect the rights and welfare of these subjects.

3.4. IRB Membership Requirements

The requirements for IRB membership are addressed in both the FDA and HHS regulations and are identical. The requirements are detailed below [8-10].

- Each IRB shall have at least five members with varying backgrounds to promote complete and adequate review of research activities commonly conducted by the institution. The IRB shall be sufficiently qualified through the experience and expertise of its members, and the diversity of the members, including consideration of race, gender, and cultural backgrounds and sensitivity to such issues as community attitudes, to promote respect for its advice and counsel in safeguarding the rights and welfare of human subjects. In addition to possessing the professional competence necessary to review specific research activities, the IRB shall be able to ascertain the acceptability of proposed research in terms of institutional commitments and regulations, applicable law, and standards of professional conduct and practice. The IRB shall therefore include persons knowledgeable in these areas. If an IRB regularly reviews research that involves a vulnerable category of subjects, such as children, prisoners, pregnant women, or handicapped or mentally disabled persons, consideration shall be given to the inclusion of one or more individuals who are knowledgeable about and experienced in working with these subjects.

- Every nondiscriminatory effort will be made to ensure that no IRB consists entirely of men or entirely of women, including the institution's consideration of qualified persons of both sexes, so long as no selection is made to the IRB on the basis of gender. No IRB may consist entirely of members of one profession.
- Each IRB shall include at least one member whose primary concerns are in scientific areas and at least one member whose primary concerns are in nonscientific areas.
- Each IRB shall include at least one member who is not otherwise affiliated with the institution and who is not part of the immediate family of a person who is affiliated with the institution.
- No IRB may have a member participate in the IRB's initial or continuing review of any project in which the member has a conflicting interest, except to provide information requested by the IRB.
- An IRB may, in its discretion, invite individuals with competence in special areas to assist in the review of issues which require expertise beyond or in addition to that available on the IRB. These individuals may not vote with the IRB.

4. Federalwide Assurance (FWA)

4.1. Defining FWA

The HHS human subject protection regulations and policies require that any institution engaged in non-exempt human subjects research conducted or supported by HHS must submit a written assurance of compliance to the Office for Human Research protections (OHRP) (**Box 8-3**). The Federalwide Assurance (FWA) is the only type of assurance of compliance accepted and approved by OHRP. Through the FWA, an institution commits to HHS that it will comply with the requirements in the HHS Protection of Human Subjects regulations at 45CFR46.

FWAs also are approved by OHRP for federalwide use, which means that other federal departments and agencies that have adopted the Federal Policy for the Protection of Human Subjects (also known as the Common Rule) may rely on the FWA for the research that they conduct or support. Institutions engaging in research conducted or supported by non-HHS federal departments or agencies should consult with the sponsoring department or agency for guidance regarding whether the FWA is appropriate for the research in question.

Box 8-3. When is an institute considered to be engaged in human subjects research?

According to HHS regulations, in general, an institution is considered to be engaged in human subjects research when its employees or agents (i) obtain data about living individuals for research purposes through intervention or interaction with them, (ii) obtain individually identifiable private information for research purposes, or (iii) obtain the informed consent of human subjects.

Employees and agents, including students, are individuals performing institutionally designated activities and acting on behalf of the institution or exercising institutional authority or responsibility.

4.2. Key Features of FWA

The key features of FWA are the following [13]:

- The identifying information for the institution filing the FWA, the human protections administrator (or a reliable point of contact) at the institution, and the institutional official signing the FWA.
- A list of the institution's legal components that operate under different names that will be covered by the FWA and the city and state or country where the component is located.
- A statement of ethical principles to be followed in protecting human subjects of research.
- An applicability statement indicating that the FWA applies whenever the institution becomes engaged in human subjects research (see **Box 8-3** for definition) conducted or supported by any U.S. federal department or agency that has adopted the U.S. Federal Policy for Protection of Human Subjects (known as the Common Rule), unless the research is exempt from Common Rule requirements or a Common Rule agency or department determines the research will be conducted under a separate assurance of compliance. U.S. institutions may voluntarily extend the Common Rule or 45CFR46 to all research conducted by the institution regardless of the source of support.
- An assurance of compliance indicating that the institution will comply with the terms of the FWA. An institution outside of the U.S. also must assure that it will comply with one or more procedural standards whenever it engages in research covered by the FWA. A list of six such standards is included or an institution can submit another standard for the protection of human subjects that is recognized by the Common Rule departments and agencies.
- The designation of all internal IRBs that will review the research covered by the FWA. If the institution has no internal IRB, it must designate the external IRB that reviews all

research covered by the FWA. If the institution relies upon multiple external IRBs, the institution should designate the external IRB that reviews the largest percentage of the research covered by the FWA. All IRBs designated on an institution's FWA must be registered with the Office for Human Research protections (OHRP) before the FWA can be approved. It should be noted that all IRBs reviewing research covered by an institution's FWA must be registered whether or not they are designated on the institution's FWA.

- Whenever the Institution relies upon an IRB operated by another institution or organization for review of research covered by the FWA, the institution must ensure that this arrangement is documented by a written agreement between the institution and the other organization or institution operating the IRB. The agreement must outline their relationship and include a commitment that the IRB will adhere to the requirements of the Institution's FWA. OHRP's sample IRB Authorization Agreement may be used for this purpose, or the parties involved may develop their own agreement. This agreement must be kept on file at both institutions/ organizations and made available to OHRP or any U.S. federal department or agency conducting or supporting research covered by the FWA upon request.

5. Glossary Pertaining to the Protection of Human Subjects in Research

Protection of human subjects in research is a broad issue facing the scientific community and the society as a whole. The main aim of this chapter is to provide an overview on basic ethical principles and regulations regarding the protection of human subjects in biomedical research. It is hoped that such a concise coverage will provide enough background information to the readers, especially those who are decision-making, in developing the strategic plan for institutional biomedical research involving human subjects. To help the readers gain a better understanding of the principles of protection of human subjects in research, additional resources of bioethics and commonly encountered terms related to the protection of human subjects in biomedical research are listed in **Table 8-2**.

Table 8-2. Additional resources of bioethics and commonly encountered terms pertinent to the protection of human subjects in biomedical research

Resource or Term	Website/Description or Definition
NIH's Bioethics Resources on the Web	http://bioethics.od.nih.gov This NIH website provides a compilation of web links to information of potential use and interest to diverse audiences, including: • Biomedical and behavioral researchers • Health care professionals • Research participants • Patients and patient advocates • Students and faculty in different disciplines • The general public

RESOURCE OR TERM	WEBSITE/DESCRIPTION OR DEFINITION
	With the aim of helping individuals and groups explore the vast array of issues in bioethics, the information accessible through this portal spans a wide range of topics, including human subjects and animal research, Institutional Review Boards (IRBs; also known as research ethics committees), clinical ethics, international research ethics issues, genetics, and neuroethics, among others. Teaching materials (including case studies and curricula), conferences and events, journals in the field, and academic centers and institutes in bioethics are just a few types of the information accessible through this site, which also provides links to information about bioethics initiatives at the institutes and centers of NIH and other government agencies. For beginning students, as well as more experienced professionals and laypeople, Bioethics Resources on the Web offers easy access to key documents and references in the field.

CITI (Collaborative Institutional Training Initiative)	https://www.citiprogram.org
	The Collaborative Institutional Training Initiative (CITI) was founded in 2000 and offers courses to participating institutions and organizations around the world by subscription. The online course topics include the following:
	• Basic Courses in the Protection of Human Research Subjects
	    • Biomedical Focus
	    • Social and Behavioral Focus
	    • Refresher Courses
	• Good Clinical Practice Courses
	    • Good Clinical Practice Course Drugs and Devices (FDA-Focused)[a]
	    • Good Clinical Practice Course for Clinical Trials Involving Drugs (ICH[b]-Focused)
	    • Good Clinical Practice Course for Clinical Trials Involving Devices
	• Health Information Privacy and Security Course (HIPS) including HITECH[c]
	• Laboratory Animal Research Courses For Investigators and IACUC Members
	    • Animal Model Specific Courses.
	• Responsible Conduct of Research (RCR) Courses
	    • Biomedical Research Focus
	    • Social, Education, Behavioral Research Focus
	    • Physical Sciences Research Focus
	    • Scholarly Activities in The Humanities Focus
	    • RCR for Engineers
	    • RCR for Science Administrators
	• Bio-Safety and Bio-Security
	
	[a]see Section 3.4.4 and Table 1-10 of Chapter 1 for definition of Good Clinical Practice
	[b]ICH stands for International Conference on Harmonization.
	[c]HITECH stands for the Health Information Technology for Economic and Clinical Health (HITECH) Act, enacted as part of the American Recovery and Reinvestment Act of 2009, and signed into law on February 17, 2009, to promote the adoption and meaningful use of health information technology.

Table 8-2. (Continued)

RESOURCE OR TERM	WEBSITE/DESCRIPTION OR DEFINITION
ANPRM for Revision to Common Rule	The Office of the Secretary of the Department of Health and Human Services (HHS) in coordination with the Office of Science and Technology Policy (OSTP) issued in 2011 the advance notice of proposed rulemaking (ANPRM) to request comments on how current regulations for protecting human subjects who participate in research might be modernized and revised to be more effective. The ANPRM seeks comments on how to better protect human subjects who are involved in research, while facilitating valuable research and reducing burden, delay, and ambiguity for investigators.
Assent	Assent, as it pertains to the protection of children as subjects in research, is defined in 45CFR46.402, Subpart D as a child's affirmative agreement to participate in research. Mere failure to object should not, absent of affirmative agreement, be construed as assent.
Biobank	An organized collection of human biospecimens (blood, tissues, and other biological materials) and associated information stored for medical research purposes.
Children	There are separate definitions for the protection of children in the HHS regulations and for the inclusion of children in the NIH policy, as follows: • From the HHS regulations (45CFR46.402, Subpart D)- persons who have not attained the legal age for consent to treatments or procedures involved in the research, under the applicable law of the jurisdiction in which the research will be conducted. • From NIH Policy (NIH Policy and Guidelines on the Inclusion of Children as Participants in Research)- an individual under the age of 21 years. Notably, the age limit for children in the NIH Policy is different from that for the legal age of adult, which is 18 years in the United States.
Coded	Coded, as it pertains to private information or human biological specimens, is defined in the OHRP Guidance on Research Involving Coded Private Information or Biological Specimens as: • having identifying information (such as name or social security number) that would enable the investigator to readily ascertain the identity of the individual to whom the private information or specimens pertain, and includes replacing identifying information with a number, letter, symbol, or combination thereof (i.e., the code); and • a key to decipher the code exists, enabling linkage of the identifying information to the private information or specimens.
Confidentiality	The manner of treating private information, which has been disclosed by the individual subject of the information to a particular person or persons for a specific purpose, such that further disclosure of the information will not be allowed to occur without authorization.
Exemption	The Common Rule codified in 45CFR46.101(b) specifies that research activities may be exempt from the policy if human subjects involvement is limited to one of the listed 6 categories, including studies involving the collection or study of existing data when those data either are publicly available or are not personally identifiable. As detailed in the text, however, some IRBs request to see the studies that may be exempt and open files on them for tracking.

Resource or Term	Website/Description or Definition
Expedited Review	The Common Rule codified in 45CFR46.110 specifies that research activities may be eligible for expedited review if the protocol involves only minimal risk or a previously reviewed protocol is receiving modifications that are only minor. Expedited review is carried out by the IRB Chair or by one or more experienced reviewers designated by the chair. Such expedited reviews have the force of full reviews, except that if the protocol is found not acceptable, then it must receive review by the full committee; the chair or designee alone cannot reject a proposal.
Individually Identifiable	It is defined in 45CFR46.102 within the Human Subject definition as the identity of the subject is or may be readily ascertained by the investigator or readily associated with the information. In addition, OHRP generally considers private information or specimens to be individually identifiable when they can be linked to specific individuals by the investigator(s) either directly or indirectly through coding systems (according to the Guidance on Research Involving Coded Private Information or Biological Specimens).
Informed Consent	A person's voluntary agreement, based upon adequate knowledge and understanding, to participate in human subjects research.
Intervention	Intervention, as it pertains to research involving human subjects is defined in 45CFR46.102 within the Human Subject definition as: • physical procedures by which data are gathered (for example, venipuncture) and • manipulations of the subject or the subject's environment that are performed for research purposes.
Investigator	It is defined in the OHRP Guidance on Research Involving Coded Private Information or Biological Specimens as anyone involved in conducting the research. Individuals who provide coded information or specimens and collaborate on other activities related to the conduct of the research with the investigators who receive such information or specimens, are considered to be involved in the research.
Legally Authorized Representative	Legally Authorized Representative, as it pertains to decisionally incapacitated individuals, is defined in 45CFR46.102, Subpart A as an individual or judicial or other body authorized under applicable law to consent on behalf of a prospective subject to the subject's participation in the procedure(s) involved in the research.
Minimal Risk	The HHS regulations contain two definitions for minimal risk, as follows: • A general definition found in Subpart A (45CFR46.102)- the probability and magnitude of harm or discomfort anticipated in the research are not greater in and of themselves than those ordinarily encountered in daily life or during the performance of routine physical or psychological examinations or tests. • A specific definition for research involving prisoners found in Subpart C (45CFR46.303)- the probability and magnitude of physical or psychological harm that is normally encountered in the daily lives, or in the routine medical, dental, or psychological examination of healthy persons.
Obtain	Obtain, as it pertains to research involving human subjects, is defined in the OHRP Guidance on Research Involving Coded Private Information or Biological Specimens as to receive or access identifiable private information or identifiable specimens for research purposes and includes an investigator's use, study, or analysis for research purposes of

Table 8-2. (Continued)

Resource or Term	Website/Description or Definition
	identifiable private information or identifiable specimens already in the possession of the investigator.
Permission	Permission, as it pertains to the protection of children as subjects in research, is defined in 45CFR46.402, Subpart D as the agreement of parent(s) or guardian for ("to" in the regulations) the participation of their child or ward in research.
Private Information	Private Information, as it pertains to research involving human subjects, is defined in 45CFR46.102 within the Human Subject definition as: • information about behavior that occurs in a context in which an individual can reasonably expect that no observation or recording is taking place, and • information which has been provided for specific purposes by an individual and which the individual can reasonably expect will not be made public (for example, a medical record).

Conclusion and Perspectives

Over the past several decades, efforts have been taken to ensure the protection of human subjects in research in the United States. Among the most notable achievements in this regard are the introduction of the basic ethical principles in the Belmont Report [3] and the publication of the Federal Policy for the Protection of Human Subjects or the "Common Rule" in 1991 [4] along with the enforcement of IRB by both Department of Health and Human Services (HHS) and FDA (note: FDA is an agent under HHS; FDA has its own regulations on human subjects research due to its unique role in promoting health by ensuring the safety of food and drugs) [8-10]. Similarly, international guidelines [14] for biomedical research involving human subjects have been published to provide a framework for guiding responsible human subjects research in international community. The concept of IRB has now been becoming common practice not only in the United States, but also globally in many developed and developing nations [15-19].

Although it plays an essential role in assuring the protection of human subjects in research, the operation of IRB is also facing increasing challenges. For instance, IRB is set up to address regulatory and compliance issues involved in research, but not the broad social implications of research protocols [20, 21]. It has been argued that an awareness of the social context can improve the research design process [20, 21]. Such awareness would enhance subject satisfaction and cooperation, increase community acceptance, and improve data quality. Similarly, it is argued that due to the introduction of significant changes to the research landscape the Common Rule published more than 20 years ago needs revision and modernization to enhance protection while simultaneously eliminate unreasonable burdens [22-25]. In this context, emerging ethical issues from human stem cell research as well as biobanking over the past few years warrant the establishment of effective and efficient policies and regulations [26-30]. Likewise, effective measures also need to be developed to encourage participation in biomedical research to make the society significantly healthier and longer lived [31, 32].

References

[1] Steneck NH. ORI Introduction to the Responsible Conduct of Research (2007). http://ori.hhs.gov/documents/rcrintro.pdf (accessed December 12, 2012).

[2] Levine RS, Williams JC, Kilbourne BA, Juarez PD. Tuskegee redux: evolution of legal mandates for human experimentation. *J Health Care Poor Underserved.* 2012; 23(4 Suppl):104-25.

[3] The National Commission for the Protection of Human Subjects of Biomedical and Behavioral Research the Belmont Report: Ethical Principles and Guidelines for the Protection of Human Subjects of Research (1978). http://videocast.nih.gov/pdf/ohrp_belmont_report.pdf (accessed December 12, 2012).

[4] DHHS. Federal Policy for the Protection of Human Subjects ('Common Rule'). http://www.hhs.gov/ohrp/humansubjects/commonrule/ (accessed December 12, 2012).

[5] Office for Human Research Protections (OHRP). http://www.hhs.gov/ohrp/ (accessed December 12, 2012).

[6] Office for Human Research Protections (OHRP). Guidance on Engagement of Institutions in Human Subjects Research. http://www.hhs.gov/ohrp/policy/engage08.html (accessed December 12, 2012).

[7] Office for Human Research Protections (OHRP). Regulations. http://www.hhs.gov/ohrp/humansubjects /index.html (accessed December 12, 2012).

[8] FDA. Comparison of FDA and HHS human subject protection regulations. http://www.fda.gov/ScienceResearch/SpecialTopics/RunningClinicalTrials/educationalmaterials/ucm112910.htm (accessed December 12, 2012).

[9] FDA. CFR - Code of Federal Regulations Title 21 part 56. http://www.accessdata.fda.gov/scripts/cdrh/cfdocs/cfcfr/cfrsearch.cfm?cfrpart=56 (accessed December 12, 2012).

[10] DHHS. Code of Federal Regulations title 45 part 46. http://www.hhs.gov/ohrp/humansubjects/guidance/45cfr46.html (accessed December 12, 2012).

[11] Office for Human Research Protections (OHRP). Federalwide Assurances (FWAs). http://www.hhs.gov/ohrp/assurances/assurances/index.html (accessed December 12, 2012).

[12] Office for Human Research Protections (OHRP). When is an institution considered to be "engaged in research"? http://answers.hhs.gov/ohrp/questions/7141 (accessed December 12, 2012).

[13] Office for Human Research Protections (OHRP). What are the key features of the Federalwide Assurance (FWA)? http://answers.hhs.gov/ohrp/questions/7145 (accessed December 12, 2012).

[14] CIOMS. International Ethical Guidelines for Biomedical Research Involving Human Subjects (2002). http://www.recerca.uab.es/ceeah/docs/CIOMS.pdf (accessed December 12, 2012).

[15] Rehmann-Sutter C. Sino-European research ethics on the right path. *Nature.* 2010; 466:28.

[16] Perkovic V, Patil V, Wei L, Lv J, Petersen M, Patel A. Global randomized trials: the promise of India and China. *J Bone Joint Surg Am.* 2012; 94 Suppl 1:92-6.

[17] Li EC, Du P, Ji KZ, Wang Z. Chinese ethics review system and Chinese medicine ethical review: past, present, and future. *Chin J Integr Med.* 2011; 17:867-72.

[18] Shergold M. Guiding Good Research: Biomedical Research Ethics and Ethics Review: Observatory on Health Research Systems (2008). http://www.rand.org/content/dam/rand/pubs/documented_briefings/2008/RAND_DB536.pdf (accessed December 12, 2012).
[19] Kadam R. Proactive role for ethics committees. *Indian J Med Ethics.* 2012; 9:216.
[20] Havard M, Magnus D. Beyond the IRB: local service versus global oversight. *Am J Bioeth.* 2011; 11:1-2.
[21] Gordon JB, Levine RJ, Mazure CM, Rubin PE, Schaller BR, Young JL. Social contexts influence ethical considerations of research. *Am J Bioeth.* 2011; 11:24-30.
[22] Rice MJ. The institutional review board is an impediment to human research: the result is more animal-based research. *Philos Ethics Humanit Med.* 2011; 6:12.
[23] Lo B, Barnes M. Protecting research participants while reducing regulatory burdens. *JAMA.* 2011; 306:2260-1.
[24] Emanuel EJ, Menikoff J. Reforming the regulations governing research with human subjects. *N Engl J Med.* 2011; 365:1145-50
[25] Dresser R. Research ethics. Aligning regulations and ethics in human research. *Science.* 2012; 337:527-8.
[26] Klitzman R, Appelbaum PS. To protect human subjects, review what was done, not proposed. *Science.* 2012; 335:1576-7.
[27] Lo B, Parham L, Cedars M, Fisher S, Gates E, Giudice L, Halme DG, Hershon W, Kriegstein A, Rao R, Roberts C, Wagner R. NIH guidelines for stem cell research and gamete donors. *Science.* 2010; 327:962-3.
[28] Baker M. Biorepositories: Building better biobanks. *Nature.* 2012; 486:141-6.
[29] Master Z, Nelson E, Murdoch B, Caulfield T. Biobanks, consent and claims of consensus. *Nat Methods.* 2012; 9:885-8.
[30] Saha K, Hurlbut JB. Treat donors as partners in biobank research. *Nature.* 2011; 478:312-3.
[31] Schaefer GO, Emanuel EJ, Wertheimer A. The obligation to participate in biomedical research. *JAMA.* 2009; 302:67-72.
[32] Trinidad SB, Fullerton SM, Ludman EJ, Jarvik GP, Larson EB, Burke W. Research practice and participant preferences: the growing gulf. *Science.* 2011; 331:287-8.

Self-Assessment Questions

[1] The use of human subjects in research benefits society in many ways, from contributing to the development of new drugs and medical procedures to understanding how we think and behave. In 1991 most U.S. federal departments and agencies that conduct or support human subjects research adopted a common set of regulations for the protection of human subjects in research. This common set of regulations is referred to as:

A. The Belmont Report
B. The Common Rule
C. The Declaration of Helsinki
D. The ICH Good Clinical Practice

E. The Nuremberg Code

[2] Which of the following is not one of the basic principles outlined in the Belmont Report?

A. Beneficence
B. Double-blind masking
C. Justice
D. Respect

[3] With regard to the protection of human subjects in research, the U.S. HHS regulations 45CFR46 include four subparts. Which subpart deals with additional protections for children involved as subjects in research?

A. Subpart A
B. Subpart B
C. Subpart C
D. Subpart D

[4] Institutional Review Board (IRB), also known as institutional review committee, independent ethics committee, or ethical review board, is a committee formally designated by an institution to review, to approve the initiation of, and to conduct periodic review of, biomedical research involving human subjects. The function of IRB is defined by:

A. Both U.S. FDA and HHS regulations
B. The Belmont Report
C. The ICH Good Clinical Practice
D. The Nuremberg Code

[5] According to the U.S. HHS regulations, in general, an institution is considered to be engaged in human subjects research when its employees or agents:

A. Conduct basic biomedical research with clinical implications
B. Obtain individually identifiable private information for research purposes
C. Study a cell line derived from a human brain tumor
D. Work with primary human cardiomyocytes obtained from a commercial source

Unit IV.
Strategic Biomedical Research Planning

Chapter 9

The Three-Set-Factor Model of Strategic Biomedical Research

Abstract

Developing a strong biomedical research program relies on in place a compelling strategic research plan that addresses the vision, mission, and goals and strategies to meet the goals. This chapter introduces the three-set-factor model for establishing world-class research universities. This universal model of institutional excellence can also be used to guide the development of strategic biomedical research at both research universities and traditionally non-research-intensive institutions of higher education.

Chapter at a Glance

1. Overview

At the institutional level, establishment of a nationally or internationally-recognized research university requires in place a strategic research plan (**Box 9-1**) addressing strong leadership, compelling vision and mission statements on research, strategic research goals, and clearly articulated strategies to translate the vision and goals into concrete programs and targets [1]. Similarly, the essential elements of a strategic research plan for developing biomedical research in traditionally non-research-intensive institutions, including osteopathic medical schools (see Chapter 2 for discussion of osteopathic medicine) should include the following:

- Vision and mission statements on biomedical research
- Biomedical research goals and strategies to attain the goals

As repeatedly stated in the book, developing a compelling strategic biomedical research plan relies on a fundamental understanding of the structural components of biomedical research (which have been covered in preceding chapters) as well as the management framework. This chapter introduces the three-set-factor model for establishing world-class universities [1, 2], which can also serve as an administrative guide for strategic biomedical research planning at institutions of higher education, including those that are traditionally non-research-intensive. The subsequent Chapter 10 addresses the importance of vision and mission statements and discusses the overall strategies for developing compelling vision and mission statements on biomedical research. The goals and strategies to achieve the goals, a central component of strategic biomedical research planning, are covered in Chapter 11, the last chapter of the book.

Box 9-1. Strategic research plan: what is it and why is needed?

What is it?

- The term strategy is derived from the Greek word "στρατηγία" (strategia) meaning generalship or leading an army.
- Strategic planning may be defined in two ways: (i) a well-thought and carefully chosen plan to bring about a desired future, such as achievement of a goal or solution to a problem; and (ii) the art and science of planning and marshalling resources for their most efficient and effective use.
- Based on the above definitions, a strategic research plan can be simply defined as a strategic plan specifically addressing the research activity and development for an organization.
- As there is always an element of uncertainty about the future, a strategic research plan is more about a set of options or strategic choices rather than a fixed plan.
- A good strategic research plan should be designed for flexibility, scalability, and ongoing revision as new institutional and societal needs arise.

Why is it needed?

- Strategic research planning provides the framework for an institution and its partners to collaborate and to ensure that research resources are directed to the most crucial research needs.
- Strategic research planning helps find new resources for new and continued research, cultivate opportunities for collaboration on research with new and existing partners, and target existing institutional resources to provide the greatest benefit for the institution and the society.

- Internally, the strategic research plan can be used to guide the institution's research priorities. In this regard, the strategic research plan becomes a compass which ensures the research projects are aligned with the institution's overall strategic direction and direct resources to higher priority research.
- Externally, the strategic research plan can be used to communicate the institutional research priorities with outside parties to help find partners, develop collaborative research, and seek funding.
- For institutions of higher education, a compelling strategic research plan will help recruit and retain promising faculty researchers and improve both graduate and undergraduate research programs as well as the education enterprise as a whole.

2. The Three-Set-Factor Model of Strategic Biomedical Research

The three-set-factor model for developing world-class research universities has been discussed in the literature [1, 2]. This section is intended to provide an overview of this model in a context of developing strategic biomedical research. The three-set-factor model emphasizes the essentiality of three intertwined factors, i.e., talent, resources, and governance, to the establishment of world-class research institutions of higher education (**Figure 9-1**). Indeed, the superior outcomes of world-class universities- highly sought graduates, cutting-edge research, and dynamic knowledge and technology transfer- can be attributed essentially to the three complementary sets of factors [1, 2]:

(1) A high concentration of talent (faculty, students, and supportive staff)
(2) Abundant resources to offer a rich learning environment and conduct advanced research (public budget resources, donations, endowment revenues, tuition fees, and research grants)
(3) Favorable governance features that encourage strategic vision, innovation, culture of excellence, respect, autonomy, and flexibility

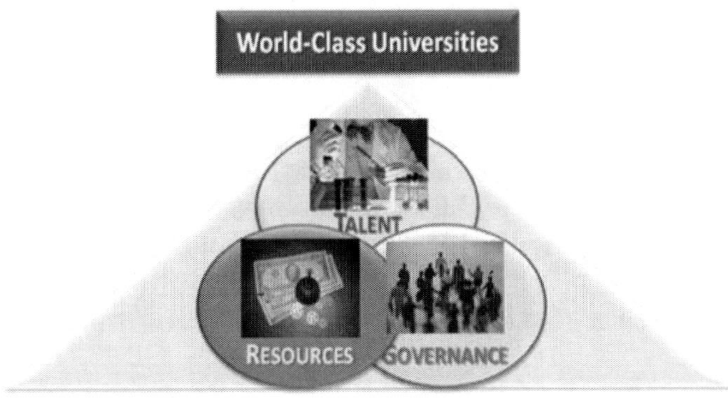

Figure 9-1. Schematic illustration of the three-set-factor model for establishing world-class universities. This universal model can also be applied to guide the development of strategic biomedical research at institutions of higher education, including those that are traditionally non-research-intensive.

2.1. Talent

The first and perhaps foremost determinant of institutional excellence is the presence of a critical mass of top students and outstanding faculty which are collectively known as academic talent. Talent refers to a group of persons with special capacity for achievement or success. Academic talent refers to those working at academic settings, and includes primarily faculty and students among others.

A key factor in building a top biomedical research institution is the ability to attract, recruit, and retain leading professors and researchers. Building a strong mass of academic talent is not only about attracting experienced professors and researchers, but also about reaching a good balance between academics at the peak of their career and young researchers with promising prospects [1, 2]. In this regard, for newly established osteopathic medical schools, one of the faculty hiring priorities should be placed on recruiting young, motivated scholars with high enthusiasm for and strong commitment to both research and teaching excellence (see Chapter 11 for more discussion in this regard).

2.2. Resources

Emerging research institutes need to be well resourced to progress rapidly (see **Box 9-2** for defining resources). Abundant funding is indispensible not only for setting up first-rate facilities and an appropriate physical infrastructure, but also for attracting and retaining high level academics. In addition, being well resourced in the early days of a new institution is not sufficient for building a top-notch research institution. Funding must be sustained over the long term.

Successful research institutions diversify their funding resources including allocation of in-house fund and donations, as well as competing for private and public research grants [1, 2]. In times of budget cuts of government funds and increasing competition for public research funding in the United States [4, 5], a sustained source of institutional intramural funding is essential for establishing a competitive research program that will eventually prosper and become substantially supported by extramural funding through attracting and retaining productive researchers [2, 5]. This is particularly true for newly established medical institutions, including osteopathic medical schools. World-class resources attract world-class scholars, who in turn attract world-class research colleagues and world-class students [2, 5, 6]. Together they create a rich educational environment that fosters innovation and intellectual excitement, which inevitably improve the institution's prestige and value and help attract extramural funding. In this context, the return of an institution's investment in research is multiplicative, and cannot (and should not) be judged solely by the pure number of extramural grant dollars that the researchers have brought or will bring to the institution.

Box 9-2: Resources: definition and characteristics [3]

Definition:

The term "resource" usually refers to asset, material, or capital which can be used to accomplish a goal.

The concept of resources has been applied in diverse realms, including biomedical research. It is linked to the concepts of competition, sustainability, conservation, and stewardship. In application to biomedical research, resource allocation through resource management is crucial for achieving research excellence. This is partly determined by the major characteristics of resources.

Three main characteristics of resources:

- Utility
- Limited availability
- Potential for consumption and depletion

2.3. Governance

Recently the terms "governance" and "good governance" are being increasingly used in the literature related to the development of world-class universities. Governance refers to the process of decision-making and the process by which decisions are implemented (or not implemented). Governance can be used in several contexts such as institutional governance, corporate governance, international governance, national governance, and local governance. Good governance has 8 major characteristics, which are also known as the eight principles of good governance (**Figure 9-2**) [7, 8]:

1. It is participatory: Good institutional governance encourages the participation of faculty, staff, and students. The talent should be part of the institutional projects rather than pure listeners to the top administrators.
2. It is consensus oriented: Good institutional governance requires mediation of the different interests of the faculty, staff, students, and administrators among other members in the institution to reach a broad consensus across the institution on what is in the best interest of the whole institution and how this can be achieved.
3. It is accountable: Accountability (an obligation or willingness to accept responsibility or to account for one's actions) is a key requirement of good institutional governance. The institution must be accountable to its stakeholders and those who will be affected by its decisions or actions. Accountability cannot be enforced without transparency and following the rule of law.
4. It is transparent: Transparency means that decisions taken and their enforcement are done in a manner that follows rules and regulations. It also means that information is freely available and directly accessible to those who will be affected by such decisions and their enforcement. It also means that enough information is provided and that it is provided in easily understandable forms and media.
5. It is responsive: Good institutional governance requires that institution or its leadership tries to serve all its stakeholders within a reasonable timeframe. Being responsive also means being open-minded and willing to listen to different opinions and take constructive suggestions.
6. It is effective and efficient: Good governance means that processes and institutions produce results that meet the needs of society while making the best use of resources

at their disposal. The concept of efficiency in the context of good governance also covers the sustainable use of natural resources and the protection of the environment.
7. It is equitable and inclusive: Good institutional governance ensures that all its members feel that they have a stake in it and do not feel being excluded.
8. It follows the rule of law: Good governance requires fair legal frameworks that are enforced impartially. It also requires full protection of human rights, particularly those of minorities. Impartial enforcement of laws requires an independent judiciary and an impartial and incorruptible police force. Following the rule of law is indispensible for good governance at the institutional level.

The concept of good governance has emerged as an essential part of sustainable development at the level of society [9]. Good governance is also crucial for the continued success of institutions of higher education through ensuring sustainable resources and talent retention [1, 10]. In contrast, bad governance lacks the above characteristics and is being increasingly regarded as one of the root causes of failure in developing a nationally or internationally recognized institution of higher education. It is also one of the root causes for the degradation of once highly-regarded institutions. In this context, it is important for the senior leadership to appreciate the essentiality of good governance to the survival and prosperity of an institution of higher education as a whole, as well as to the establishment and progress of any individual institutional programs.

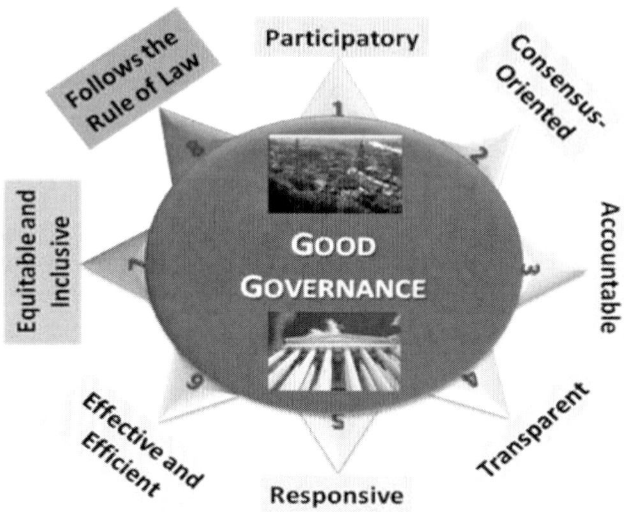

Figure 9-2. Schematic illustration of good governance. Good governance is essential not only to the development and prosperity of society, but also to the survival and progress of individual institutions of higher education.

Conclusion and Perspectives

Case studies of successful research institutions show that an appropriate regulatory framework, strong and inspiring leadership, and adequate management significantly influence the research institutions to prosper. One of the ways in which purposeful leadership manifests itself is through the talent of articulating an enticing vision for the future of the institution to all its stakeholders. Good salaries are not enough to attract and motivate high-performing academics; talent faculty members must feel that they are part of a significant project with strong support from the top administration as well as research offices and committees (see Chapters 6-8 for discussion of research offices and committees) to ensure their full commitment toward the construction or renewal of the institution [1]. Indeed, as shown in **Figure 9-3**, aligning the three sets of elements- concentration of talent, sufficient funding, and great governance- determines the research performance and consequently the improvement of prestige and value of the institutions of higher education, including osteopathic medical schools affiliated with traditionally non-research-intensive universities.

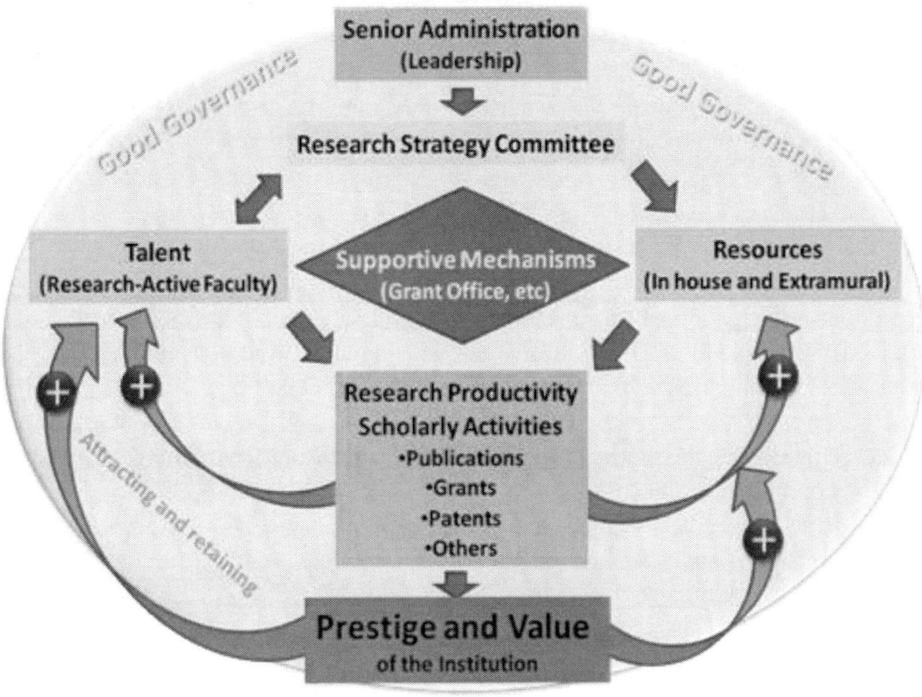

Figure 9-3. Research-dominant pathways to institutional excellence in the context of aligning the three sets of factors of a high concentration of talent, abundant resources, and good governance.

References

[1] Altbach PG, Salmi J. The Road to Academic Excellence: The Making of World-Class Research Universities (Directions in Development) (2011), the World Bank, Washington, D.C.

[2] Salmi J. Attracting talent in a global academic world: how emerging research universities can benefit from brain circulation. The Acaedmic Executive Brief. 2012; 2(1);1-5. http://academicexecutives.elsevier.com/sites/default/files/AEB_2.1_Salmi.pdf (accessed on December 15, 2012).

[3] Wikipedia. Resource. http://en.wikipedia.org/wiki/Resource (accessed on December 15, 2012).
[4] Moses H 3rd, Dorsey ER. Biomedical research in an age of austerity. *JAMA*. 2012; 308:2341-2.
[5] Sun GH, Steinberg JD, Jagsi R. The calculus of national medical research policy: the United States versus Asia. *N Engl J Med*. 2012; 367:687-90.
[6] Qiu J. China targets top talent from overseas. *Nature*. 2009; 457:522.
[7] Toward good governance. http://reform.gov.bb/page/GOOD_GOVERNANCE.pdf (accessed on December 15, 2012).
[8] United Nations Economic and Social Commission for Asia and the Pacific. What is good governance? http://www.unescap.org/pdd/prs/projectactivities/ongoing/gg /governance.pdf (accessed on December 15, 2012).
[9] Hurlbut JB, Robert JS. Good governance connects science and society. *J Policy Anal Manage*. 2012; 31:722-6.
[10] Hénard F, Mitterle A. Governance and quality guidelines in higher education: a review of governance arrangements and quality assurance guidelines. http://www.oecd.org/edu /imhe/46064461.pdf (accessed on December 15, 2012).

Self-Assessment Questions

[1] The three-set-factor model for developing world-class research universities has recently received much attention among academics and university administrators. This model emphasizes the importance of three intertwined factors, including:

A. Grant office, Institutional Review Board, and Institutional Animal Care and Use Committee
B. Resource, leadership, and students
C. Talent, resources, and governance
D. Talent, resource, and Institutional Review Board
E. Talent, students, and grant office

[2] The concept of Good Governance has emerged as an essential part of sustainable development at the level of society. Good Governance is also crucial for the continued success of institutions of higher education through ensuring sustainable resources and talent retention. In Good Governance, "Transparency" means:

A. Being open-minded and willing to listen to different opinions and take constructive suggestions
B. That information is freely available and directly accessible to those who will be affected by any decisions and their enforcement
C. That processes produce results that meet the needs of society while making the best use of resources at their disposal
D. That talent should be part of the institutional projects rather than pure listeners to the top administrators

[3] Strategic planning is a well-thought and carefully chosen plan to bring about a desired future, such as achievement of a goal or solution to a problem. In developing a biomedical research program at a traditionally non-research-intensive institution, Professor S.T. lists several reasons for having a strategic biomedical research planning in place first before mobilizing the institutional resources to build the research program. Which of the following is most unlikely to be on the list?

A. It can be used to communicate the institutional research priorities with outside parties to help find partners, develop collaborative research, and seek funding.
B. It can be used to guide the institution's research priorities and to ensure that the research projects are aligned with the institution's overall strategic direction.
C. It is the only way to ensure the full return of the institutional investment in biomedical research.
D. It provides the framework for an institution and its partners to collaborate and to ensure that research resources are directed to the most crucial research needs.

[4] In Good Governance, "Consensus-Orientated" means:

A. A high concentration of talent including faculty, students, and supportive staff
B. A strategic plan specifically addressing the research activity and development for an institution
C. Information being freely available and directly accessible to those who will be affected by any new decisions and their enforcement
D. Mediation of the different interests of the faculty, staff, students, and administrators among other members in the institution to reach a broad agreement across the institution on what is in the best interest of the whole institution and how this can be achieved

[5] Which of the following is most unlikely to be included in a strategic biomedical research plan?

A. A mission statement
B. A team-based learning model
C. A vision statement
D. The research goals
E. The 3-set-factor model

Chapter 10

Vision and Mission Statements on Strategic Biomedical Research

Abstract

Vision and mission statements are vital elements of strategic planning, including that for developing biomedical research at institutions of higher education. This chapter first defines vision and mission statements and then discusses ways to develop effective vision and mission statements on strategic biomedical research. The chapter focuses on addressing three key areas of developing a compelling mission statement on strategic biomedical research at institutions of higher education. These are: who we are, what we do, and why we do it.

Chapter at a Glance

1. Overview
2. Defining Vision and Mission Statements
3. Developing an Effective Vision Statement on Strategic Biomedical Research
   3.1. General Considerations
   3.2. Strategies for Developing an Effective Vision Statement on Strategic Biomedical Research
4. Developing an Effective Mission Statement on Strategic Biomedical Research
   4.1. The Four Steps
   4.2. The Three Key Areas
5. Conclusion and Perspectives
6. References
7. Self-Assessment Questions

1. Overview

In developing biomedical research at institutions of higher education, especially at newly established institutions, one of the first things the leadership should determine is what the strategic goals and objectives are. A good way to plan for achieving the goals and objectives is by creating a vision statement and a mission statement on biomedical research. This chapter discusses how to create compelling vision and mission statements with regard to developing strategic biomedical research.

2. Defining Vision and Mission Statements

A vision statement sets forth the direction where the institution is headed [1-3]. It concentrates on the future or what the institution intends to be and serves as a source of inspiration. The key features of a compelling vision statement include the following.

- Clarity and lack of ambiguity
- Memorable and engaging expression
- Description of a bright future
- Realistic aspirations, achievable
- Alignment with organizational values and culture

On the other hand, a mission statement tells what the institution currently does [1-3]. It concentrates on present leading to the future. Its prime function is to list the broad goals for which the institution is formed and define the key measures to achieve success. A good mission statement has the following features.

- It addresses the purpose of the institution, how to do the business, and what principles or beliefs guide the business.
- It is short enough to remember and easily communicate, and strong enough to inspire.
- It expresses the institution's purpose in a way that inspires support and ongoing commitment.
- It motivates those who are connected to the institution.

3. Developing an Effective Vision Statement on Strategic Biomedical Research

3.1. General Considerations

A compelling strategic vision statement on research is essential for developing world-class universities. Nationally and internationally well-recognized universities are exclusively also top research institutions with strong commitment to research excellence. For traditionally non-research-intensive institutions, a clear and compelling vision on research is also an indispensible component of the research strategic plan.

A clear, inspiring vision transforms an institution, propels the institution forward, grows the institution, and encourages it to expand to its greatest potential and realize the dream. It is important that the vision statement on biomedical research is consistent with the overall vision statement of the entire institution or university.

3.2. Strategies for Developing an Effective Vision Statement on Strategic Biomedical Research

There is no fixed formula for creating a vision statement. It can be a single sentence long or occupy two pages. While there is no universal format for creating a vision statement, analysis of the vision statements of successful organizations does point to several effective ways of creating a compelling vision statement on biomedical research at institutions of higher education, especially those traditionally not research-intensive. Some of the strategies for developing an effective vision statement on institutional biomedical research are outlined below [1-3].

- An effective vision statement on biomedical research describes the outcomes that are five to ten years away, although some even further out. The vision statement should describe the best possible outcomes in the context of the impact of biomedical research on both society and the institution's prestige and value (see Chapter 2 for essentiality of biomedical research).
- Remember, the purpose of the vision statement is to inspire, energize, motivate, and stimulate the creativity of the institutional talent (including faculty, students, and supportive staff) (see Chapter 9 for discussion of academic talent), and not to serve as a measuring stick for success. That is the job of goals and objectives (see Chapter 11 for discussion of strategic goals of biomedical research).
- Using strong, but realistic phases to describe the vision statement on institutional biomedical research. Take for instance the vision statement of Johns Hopkins Medicine of "providing a diverse and inclusive environment that fosters intellectual discovery, creates and transmits innovative knowledge, improves human health, and provides medical leadership to the world" [4]. "Creates and transmits innovative knowledge", "provides leadership to the world" are powerful phases used in the vision statement.
- If possible, try to summarize the vision using powerful, but realistic phases in the first paragraph of the vision statement. Take for instance the first sentence of the vision statement of Penn Medicine of "Penn Medicine is committed to remain a world-leading institution in three equally valued and inter-related missions of patient care, education, and research" [5].
- Similarly, a vision statement on biomedical research in a traditionally non-research-intensive institution, such as an osteopathic medical school may begin with "to become the best osteopathic medical school in the nation by developing competitive biomedical research...".

4. Developing an Effective Mission Statement on Strategic Biomedical Research

4.1. The Four Steps

An effective mission statement has a real impact on an institution and its members. The mission statement guides the actions of the institution, spells out its overall goals and objectives (see **Box 10-1** for definitions of goals and objectives), provides a path, and guides decision-making. It provides the framework or context within which the institution's strategies are formulated. The basic steps in creating an effective mission statement have been extensively described in the business literature and may include the following (**Figure 10-1**) [2].

- Step 1: To create an effective mission statement on biomedical research, first identify the institution's "winning idea" or most notable strength. The "winning idea" or strength will make your institution stand out from its peer institutions.
- Step 2: Next identify the key measures of your institution's success in biomedical research. Make sure you choose the most important and effective measures. This will rely on a profound understanding of the principles of biomedical research and research structural components (see Chapters 1-8).
- Step 3: Combine your "winning idea" and success measures into tangible and measurable goals and objectives.
- Step 4: Refine the words until a concise and precise statement of the mission on institutional biomedical research is made, which clearly expresses "winning ideas", effective measures, and desired outcomes. The wording of the mission is to the extent that both the leadership and faculty can infer some order of priorities in how the strategic goals and objectives are to be met.

Box 10-1. Goals and objectives*

Goal and objective are probably the two most frequently used words in describing strategic biomedical research planning. They are, however, often confused with each other. Goal and objective both describe things that an institution may want to achieve or attain, but they differ with regard to scope, specificity, timeframe, tangibility, and measurability.

- Goals are broader than objectives in the sense that goals are general intentions and are not specific enough to be measured. Objectives are narrow and are set for certain tasks in particular.
- Goals are general whereas objectives are specific. Goals are just general intentions towards the attainment of something while objectives are precise actions for accomplishment of a specific task.
- Goals usually have a longer time-frame than objectives. Objectives are usually precise targets set for a short term. Goals may be set for a longer term but many objectives may be set within that goal.
- Goals may be intangible while objectives ought to be tangible. Goals may be directed at achieving non-measurable things while objectives ought to be targeted at getting measurable things or tasks.
- Goals may or may not be measured; however, in most cases objectives are measurable.

*modified from http://www.diffen.com/difference/Goal_vs_Objective (accessed January 7, 2013)

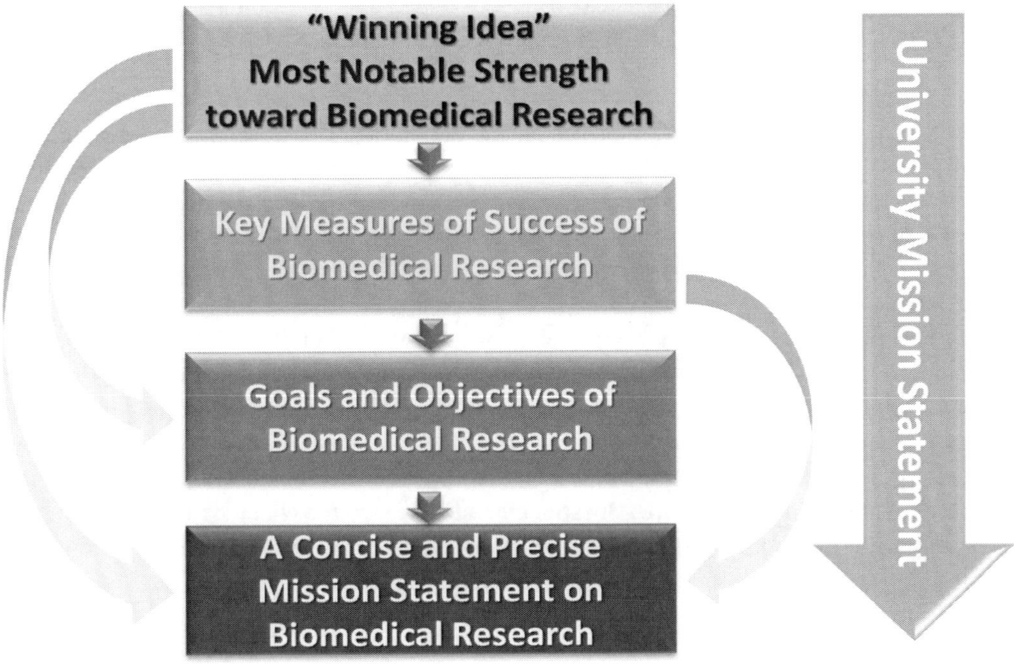

Figure 10-1. The four basic steps in developing an effective mission statement on biomedical research at institutions of higher education. Notably, although the four steps are sequential, each of the upstream steps contributes to the final creation of an effective mission statement on biomedical research. It is worth mentioning that the mission statement on biomedical research must be consistent with the overall mission statement of the university.

4.2. The Three Key Areas

While the principles of developing a mission statement for business could be applied to institutions of higher education, it is important to note the differences between for-profit and nonprofit organizations. The most fundamental difference between non-profit and for-profit organizations is the reason they exist. For-profit companies are generally founded to generate income for entrepreneurs and their employees, while nonprofits are generally founded to serve a humanitarian or environmental need. In this regard, a strong mission statement on biomedical research in institutions of higher education, especially medical schools should address three key areas [1-3]:

(1) Who we are
(2) What we do
(3) Why we do it

Since osteopathic medical schools traditionally lack emphasis on biomedical research as compared to allopathic medical institutions, this section focuses on discussing how to create effective mission statements on biomedical research in these schools. Indeed, an effective mission statement on biomedical research is critically important for advancing the prestige and value of osteopathic medical schools in the context of contribution to education and human health.

4.2.1. Who We Are

With regard to creating a mission statement on biomedical research, the meaning of "Who we are" is two-fold: (i) what the purpose of the institution is, and (ii) what the current status of the institution is.

4.2.1.1. Purpose of the Institution Determines the Mission Statement on Biomedical Research

As education is one of the central components of institutions of higher education, developing biomedical research in such organizations should be constructed in a way that can directly or indirectly benefit the education of the students. Indeed, undergraduate biomedical research has been part of the mission for most of the top medical schools as well as many traditionally non-research-intensive institutions. Undergraduate research has gained further national prominence as an effective educational strategy given the changing demographics of the students, the expanding scope of institutions providing undergraduate education, and the enhanced appreciation for active-learning strategies. Including undergraduate biomedical research in the mission statement on biomedical research will not only strengthen the education, but also benefit the institution in many ways. Some of the benefits are outlined in **Table 10-1** [6-9].

Table 10-1. A partial list of the benefits of undergraduate research at institutions of higher education [6-9]

Category	Description
Benefits to Students	• Advancing cognitive and intellectual growth, including (i) better understanding the knowledge learned in classroom, (ii) gains of additional knowledge and skills, and (iii) increased academic achievement and educational attainment. • Fostering professional growth and advancement, including (i) enhanced ability to work collaboratively with others, (ii) establishing a relationship with faculty mentors, (iii) enhanced professional credentials, and (iv) increased chance to be accepted into competitive post-baccalaureate programs. • Promoting personal growth, including (i) boosting personal confidence and stimulating curiosity, (ii) enhanced ability to learn independently and think critically, (iii) enhanced understanding of the philosophy of lifelong learning, and (iv) enhanced opportunity to become an academic role model. • Maximizing high impact learning experiences.
Benefits to Faculty	• Enhancing mentoring and teaching, including (i) to mentor high quality and engaged students, (ii) to improve classroom teaching, (iii) to be an academic role model and influence the career development of the undergraduate students, and (iv) to foster synergy between graduate and undergraduate programs. • Achieving research and scholarly accomplishments. • Integrating scholarly and teaching activities. • Enhancing faculty job satisfaction and career development.

CATEGORY	DESCRIPTION
Benefits to Institution	• Enriching the institution's curriculum. • Building a community of scholars and providing opportunities to engage with the local community. • Deepening relationships with alumni. • Fostering innovation and cross-talk, such as establishing a campus-wide undergraduate research program. • Sharing a sense of purpose and achievement, including generation of a strong voice to emphasize the importance of learning through discovery.

4.2.1.2. CURRENT STATUS OF THE INSTITUTION DETERMINES THE MISSION STATEMENT ON BIOMEDICAL RESEARCH

Osteopathic medical schools emphasize educating primary care physicians with relatively less effort and resources being devoted to create and translate new knowledge to benefit patient care (see Chapter 2 for more discussion of osteopathic medicine). The relative lack of both high quality basic and clinical research has been one of the major criticisms that osteopathic medical schools have received over the last several decades [10]. Regardlessly, as mentioned in Chapter 2, multiple osteopathic medical schools are highly-recognized as some of the best medical schools in primary care by the U.S. News & World Report. Encouragingly, over the past several years, osteopathic medicine profession has witnessed increased investment in biomedical research and the consequent development of recognized research programs in several osteopathic medical schools. Osteopathic medicine embraces and emphasizes the principles of integrative medicine and holistic health, which provide great opportunities for developing translational and clinical research (also see Section 4.2.2 below for further discussion).

4.2.2. What We Do

As noted in Chapter 2, osteopathic medicine emphasizes holistic health with the use of osteopathic manipulative treatment (OMT) and a focus on primary care being the most often cited rationales for its uniqueness. The osteopathic holistic philosophy is in line with the principles of integrative medicine, an entity widely claimed as the future of medicine and approach of patient care [11-15]. Research in osteopathic medical schools should certainly address the "osteopathic areas" including OMT, mind-body medicine, and other complementary and alternative modalities (e.g., exercise, dietary factors, and natural products, among many others) (see **Box 10-2** for difference between complimentary therapies and alternative therapies). The importance and urgency of research in the above "osteopathic areas" are evidenced by the commitment to research by United States National Institutes of Health's National Center for Complementary and Alternative Medicine (NCCAM, see **Table 10-2** for the mission and funding priorities of NCCAM) [16].

Box 10-2. Major differences between complimentary therapies and alternative therapies

Complementary therapies are adjuncts to mainstream treatment. They are safe and evidenced based-therapies that can improve the effectiveness of the mainstream therapies or reduce the adverse effects associated with them.

Complementary therapies are distinguished from alternative therapies (see below). Examples of complementary therapies include acupuncture, mind-body medicine (meditation, yoga), and physical activity.

Alternative therapies are modalities with unproven clinical effectiveness and efficacy. They might be potentially harmful. Alternative therapies are frequently promoted as substitutes for mainstream therapies. Examples of alternative therapies include certain herbal products and some dietary supplements. It is important to note that as research progresses some previously unproven alternative modalities may prove to be effective and safe for the management of certain human diseases. In this regard, many alternative modalities like complementary therapies deserve further investigation to establish their clinical effectiveness, efficacy, and the underlying mechanisms of action.

Table 10-2. NCCAM and its mission and funding areas [16]

Mission	To define, through rigorous scientific investigation, the usefulness and safety of complementary and alternative medicine interventions and their roles in improving health and health care. NCCAM achieves its mission through basic, translational ("bench-to-bedside"), and clinical research; research capacity building and training; and education and outreach programs.
Funding Areas	• Natural products • Mind-body medicine • Manipulative and body-based practices • Other complementary and alternative medicine (CAM) practices

The focus of research in osteopathic medical schools on the “osteopathic areas” doesn’t necessarily mean that researchers (both biomedical and clinical scientists) at osteopathic medical schools should not do research in other areas of medicine. In fact, research in the "non-osteopathic areas” should also be encouraged and in particular, the integrative research investigating the combined effects of “osteopathic modalities” and conventional medical treatment approaches should be strongly heartened and supported. This will greatly advance the field of integrative medicine- the future of medicine and patient care. In the author’s point of view, osteopathic medicine, its philosophy and evidence-based approaches, will greatly impact the development of integrative medicine. It can also be rationalized that osteopathic medicine along with its evidenced-based holistic approaches may become a cornerstone of future medicine and practice of patient care. **Box 10-3** outlines the basic principles of integrative medicine.

Box 10-3. Integrative medicine and its basic principles

Integrative medicine (also called integrated medicine) is healing-oriented medicine that takes account of the whole person (body, mind, and spirit), including all aspects of lifestyle. It emphasizes the therapeutic relationship and makes use of all appropriate therapies, both conventional treatment and complementary and alternative modalities for which there is some high-quality evidence of safety and effectiveness. Integrative medicine programs/departments have been established in many top medical schools in the United States [11].

The basic principles of integrative medicine include the following [17]:

- A partnership between patient and practitioner in the healing process
- Appropriate use of conventional and evidenced-based complementary/alternative methods to facilitate the body's innate healing response
- Consideration of all factors that influence health, wellness and disease, including mind, spirit, and community as well as body
- A philosophy that neither rejects conventional medicine nor accepts alternative therapies uncritically; evidence-based, patient-centered practices are the central dogma
- Recognition that good medicine should be based on good science, be inquiry driven, and be open to new paradigms
- Use of natural, effective, less-invasive interventions whenever possible
- Use of the broader concepts of promotion of health and the prevention of illness as well as the treatment of diseases
- Training of practitioners to be models of health and healing, committed to the process of self-exploration and self-development

4.2.3. Why We Do It

The last question an effective mission statement needs to address is why we do it. Specifically, why we need to develop biomedical research at the institution. The most obvious reason is to fulfill the vision on biomedical research, and as a result, to increase the visibility, prestige, and value of the institution and the purpose for which it exists. As discussed in Chapter 2, biomedical research benefits public health and social and economic development. Hence, a strong mission statement on biomedical research at an institution of higher education should emphasize the broad impact on society and human health. This broad impact largely determines the value of the institution. In this regard, a compelling mission statement on biomedical research at an osteopathic medical school may read as following: the mission of our biomedical research is to enrich the institutions' education and increase its prestige and value through supporting and conducting innovative biomedical research.

Conclusion and Perspectives

In developing strategic biomedical research, one of the first things that the leadership should do is to determine the goals and strategies to meet the goals. A good way for such planning is to develop effective vision and mission statements on biomedical research. Vision and mission statements are vital components of strategic planning to achieve success in biomedical research because they keep the institution grounded on core values and objectives. Indeed, successful research institutions, including medical schools, always have highly effective vision and mission statements to guide their research enterprises. It should be noted that the vision and mission statement on developing strategic biomedical research must also be consistent with the principles of purpose and goals and objectives set forth in the university's overall vision and mission statements. As developing effective vision and mission statements on biomedical research relies on a profound understanding of the principles of biomedical research and its research structural components (covered in Chapters 1-8), it is important that biomedical science faculty experts and institutional leadership work together to reach a consensus at the institutional level on the essentiality and goals of developing institutional biomedical research enterprise. The subsequent Chapter 11 discusses strategic biomedical research goals and strategies to attain the goals.

References

[1] Mission statement vs vision statement. http://www.diffen.com/difference/Mission_Statement_vs_Vision_Statement (accessed on December 16, 2012).

[2] Mission statements and vision statements: unleashing purpose. http://www.mindtools.com/pages/article/newLDR_90.htm (accessed on December 16, 2012).

[3] Coon DW. Creating a vision and mission statement for your organization. http://www.cwu.edu/~cel/cms/uploads/VisionMissionStatements.pdf (accessed on December 16, 2012).

[4] Johns Hopkins Medicine. Johns Hopkins Medicine vision. http://www.hopkins medicine.org/about/mission.html (accessed on December 16, 2012).

[5] Penn Medicine. Mission and vision of Penn Medicine http://www.pennmedicine.org /health-system/about/organization/mission-vision.html (accessed on December 16, 2012).

[6] Osborn, J. M. and K. K. Karukstis. The benefits of undergraduate research, scholarship, and creative activity. In: M. Boyd and J. Wesemann (Eds.), Pages 41-53, Broadening Participation in Undergraduate Research: Fostering Excellence and Enhancing the Impact. Council on Undergraduate Research, Washington, D.C., 2009. http://www.missouriwestern.edu/appliedlearning/documents/Chapter4-Final-BenefitsofUGR Osborn.pdf (accessed on December 16, 2012).

[7] Handelsman J. Dedicated to winning the future through undergraduate research. *DNA Cell Biol.* 2012; 31:891-2.

[8] Fechheimer M, Webber K, Kleiber PB. How well do undergraduate research programs promote engagement and success of students? *CBE Life Sci Educ.* 2011; 10:156-63.

[9] Simunovic F. Is there a place for medical students in research laboratories? a student's perspective. *Med Teach.* 2008; 30:875-6.

[10] Cruser dA, Dubin B, Brown SK, Bakken LL, Licciardone JC, Podawiltz AL, Bulik RJ. Biomedical research competencies for osteopathic medical students. *Osteopath Med Prim Care.* 2009; 3:10.

[11] Horrigan B, Lewis S, Abrams D, Pechura C. Integrative Medicine in America (2012). http://www.bravewell.org/content/Downlaods/IMinAm.pdf (accessed on December 16, 2012).

[12] Weisfeld V. Integrative Medicine and the Health of the Public: A Summary of the February 2009 Summit, National Academies Press, Washington D.C., 2009 http://www.nap.edu/catalog.php?record_id=12668#toc (accessed on December 16, 2012).

[13] Rayner JA, Willis K, Pirotta M. What's in a name: integrative medicine or simply good medical practice? *Fam Pract.* 2011; 28:655-60.

[14] Prasad K, Sharma V, Lackore K, Jenkins SM, Prasad A, Sood A. Use of complementary therapies in cardiovascular disease. *Am J Cardiol.* 2013; 111:339-45.
[15] Cassileth BR, Gubili J, Simon Yeung K. Integrative medicine: complementary therapies and supplements. *Nat Rev Urol.* 2009; 6:228-33.
[16] NCCAM. NCCAM Facts-at-a-glance and mission. http://nccam.nih.gov/about /ataglance (accessed on December 16, 2012).
[17] Weil A. What is integrative medicine? http://www.drweil.com/drw/u/ART02054 /Andrew-Weil-Integrative-Medicine.html (accessed on December 16, 2012).

Self-Assessment Questions

[1] Which of the following is a distinct feature of an effective vision statement as compared to a mission statement?

A. Is focusing on the future
B. Is clear
C. Is engaging
D. Is memorable

[2] The terms complementary and alternative therapies are frequently encountered in the literature and discussed by public media. Which of the following best describes complementary therapies?

A. Examples of complementary therapies include certain herbal products and some dietary supplements.
B. They are frequently promoted as substitutes for mainstream therapies.
C. They are modalities with unproven clinical effectiveness and efficacy.
D. They are safe and evidenced based-therapies that can improve the effectiveness of the mainstream therapies or reduce the adverse effects associated with them.

[3] The U.S. National Center for Complementary and Alternative Medicine (NCCAM) is part of:

A. The U.S. Center for Disease Control and Prevention
B. The U.S. Food and Drug Administration
C. The U.S. National Institutes of Health
D. The U.S. Department of Agriculture

[4] Which of the following is an example of alternative therapies?

A. Acupuncture
B. Certain herbal product
C. Mind-body medicine
D. Physical activity

[5] Integrative Medicine is considered the future of medicine and practice of patient care. Which of the following best describes Integrative Medicine?

A. Integrative medicine doesn't embrace the holistic approach.
B. Integrative medicine excludes mind-body medicine.
C. Integrative medicine is also known as alternative medicine.
D. The basic concepts of integrative medicine are embedded in osteopathic principles.

Chapter 11

Biomedical Research Goals and Strategies to Attain the Goals

Abstract

Goals and strategies to meet the goals are the core of strategic biomedical research planning at institutions of higher education. Well-defined goals and strategies to attain the goals are essential for achieving the biomedical research mission. This chapter describes five major biomedical research goals and the corresponding strategies to attain the goals. The five major goals are: (i) allocating resources and building research infrastructure to optimize resource utilization, (ii) recruiting and retaining research-active and multitasking faculty to secure institutional academic talent, (iii) defining biomedical research areas and priorities to build institutional signature research, (iv) stimulating biomedical research and scholarly productivity of faculty to maximize institutional research output and investment return, and (v) fostering interdisciplinary programs to unify university wide research enterprise.

Chapter at a Glance

1. Overview

As noted in preceding chapters, biomedical research is a main domain of the 21st century scientific research that greatly impacts public health, social and economic development, and the prestige and value of the institutions. Almost all major world-class universities conduct cutting-edge research in biomedical science, and advancing biomedical research is a top priority at these institutions. Because of its far-reaching impact, biomedical research has also received great attention at traditionally non-research-intensive institutions, including osteopathic medical schools. However, developing biomedical research is a daunting task, especially at teaching universities that lack research infrastructure and have limited resources and where the senior leadership lacks an understanding of the essentiality of biomedical research to an institution’s prestige and value. The success of biomedical research at such institutions requires the collaborative efforts of faculty, staff, and senior administrators that are guided by a compelling strategic research plan with well-defined goals and strategies. This chapter describes five major research goals and the corresponding strategies to achieve the goals. It is intended to provide a general framework for planning and execution of biomedical research activities at traditionally non-research-intensive institutions, especially osteopathic medical schools.

2. Goal One: Allocating Resources and Building Research Infrastructure to Optimize Institutional Resource Utilization

World-class research requires world-class facilities and support. The qualities of research facilities and the strength of support correlate with the excellence and impact of the institutional research activities. Although part of the research could be done through collaboration with major research universities (challenge: researchers at major research universities usually collaborate with those at similar levels of research institutions), a minimum amount of facilities and support is indispensible for developing a functional and productive research program in an osteopathic medical school. Without good facilities, it is difficult, if not impossible, to attract and retain good researchers. Listed below are some strategies to help realize goal one.

2.1. Strategy 1: To Make the Best Use of Existing Research Infrastructure and Resources

Teaching universities usually have limited resources and infrastructure devoted to cutting-edge biomedical research. Listed below are measures that may help optimize the use of existing research infrastructure and resources to maximize the research output.

- Research faculty, supportive staff, and senior administrators should work together to analyze and report on existing facilities, equipment, and supportive offices and committees across the entire campus, ranging from individual research laboratories to departmental and institutional shared core facilities. This will help generate a comprehensive list of all existing institutional resources and infrastructure. From such a list, consolidation of scattered facilities and equipment may help develop efficient institutional shared core research facilities (see Chapter 3 for discussion of core biomedical research facilities and facility management).
- To promote the efficient use of shared facilities, it is imperative to develop an online inventory and database of the existing facilities, equipment, and other research-related resources so that researchers know what is available, where it exists, and whether it is available for use. Such an online inventory and database are widely used at top research institutions in the United States, such as Johns Hopkins Medical Institutions [1] and Harvard Medical School [2].

2.2. Strategy 2: To Upgrade Existing Facilities and Purchase the Mostly Needed Equipment

Top research requires top research facilities. The success of biomedical research relies on the continued availability of sufficient institutional fund for upgrading and maintaining the existing facilities (e.g., core research laboratories and animal facilities) and for purchasing the mostly needed common equipment for cutting-edge biomedical research. As discussed in Chapter 3, the prestigious medical schools in the United States are exclusively equipped with first-rate research facilities. To maximize the output of core research facilities, efficient management is indispensible (see Section 4 of Chapter 3 for management of core facilities). In this context, for traditionally non-research-intensive institutions, it is important to establish an institutional committee to oversee the proper allocation of the limited institutional funds for establishing, optimizing, and upgrading the shared core research facilities and major equipment items.

2.3. Strategy 3: To Build New Infrastructure Dedicated to Biomedical Research

Building new research infrastructure, especially research laboratories is both time-consuming and costly (see Chapter 5 for discussion of faculty researcher independent laboratories). It is crucial to have a strategic vision on research in place to convince the stakeholders that long-term investment in biomedical research including building new research infrastructure is essential to the improvement of the prestige and value of the medical (or other health science) school and the university as a whole, and that the return of such an investment is multiplicative.

While building institutional research laboratories and other research infrastructure is usually supported by institutional budget, federal funds are also available for such a purpose. For example, as discussed in Chapter 6, the National Center for Research Resources of the National Institutes of Health (NIH) supports the Research Facilities Improvement Program (RFIP). RFIP grants are provided to public and nonprofit private biomedical institutions to expand, remodel, and renovate or alter existing research and animal facilities or construct new research and animal facilities. Improvements under this program must support basic and/or clinical biomedical and behavioral research, and they may also support research training [3].

3. Goal Two: Recruiting and Retaining Research-Active and Multitasking Faculty to Secure Institutional Academic Talent

World-class research needs world-class talent. The success of developing a research program or research institution depends on the success in recruiting and retaining talented researchers [4, 5]. In developing a research program in a newly established medical institution, especially osteopathic medical school which is primarily teaching-orientated and usually lacking competitive biomedical research, it is important to recognize the reality: best researchers are attracted and retained by best research institutions. With this in mind, it is imperative for osteopathic medical schools to recruit a significant number of faculty who are young and energetic and motivated to do both productive research and excellent teaching. In this context, it is often times seen that a good researcher is frequently also a good teacher. However, supportive mechanisms need to be in place to allow the research-active faculty (who are also committed to teaching excellence) enough time to do research in addition to providing them other necessary incentives for doing quality research (see Section 3.2 below for discussion of incentives). It is not suggested to recruit pure top researchers who are extramurally well-funded (e.g., currently supported by more than one major NIH R01) (see Table 6-2 of Chapter 6 for more information on NIH R01 grant). This is because the continued success of these individuals heavily depends on the availability of advanced institutional facilities and comprehensive institutional support, and consequently, it is difficult to retain such individuals in a traditionally non-research-intensive institution that currently lacks first-rate research facilities and corresponding supportive staff (see Chapter 3 for discussion of core biomedical research facilities, and Chapters 6-8 for description of institutional supportive entities for strategic biomedical research).

3.1. Strategy 1: To Recruit Multitasking Faculty

Because the survival and prosperity of traditionally non-research-intensive institutions depend on the excellence of faculty's teaching activities, developing biomedical research in these institutions must be based on the understanding that research cannot replace teaching or vice versa. With this in mind, it is important for such institutions to recruit faculty who have established track records in doing both excellent teaching and productive research and who are also committed to the continued success in both teaching and research as well as other institutional activities necessary for developing biomedical research. Listed below are measures that may be used to help attract the "multitasking" or "pluripotent" faculty.

- Candidate's potential for and commitment to both teaching and research excellence are used as an important criterion for hiring.
- A start package and a decent independent laboratory space are to be provided to those individuals who are committed to research excellence (see Chapter 5 for more discussion of laboratory space and start fund). The start package ($100,000-$200,000 per faculty researcher is typical for traditionally non-research-intensive institutions) is used to: (a) set up their independent laboratories, (b) purchase routine laboratory equipment, (c) develop new lines of research, and (d) generate publications and preliminary data for subsequent applications for extramural grants.
- To clearly communicate with promising candidates the strategic vision on research to convince them that institutional support will be provided, and importantly, they will be part of the institutional research and teaching enterprise, a notion that is widely appreciated in good governance (see Chapter 9 for discussion of good governance). The key message is that both teaching and research are important and one cannot be replaced by the other if the institution is deemed to improve its prestige and value via investing in biomedical research [6, 7].

3.2. Strategy 2: To Retain the Faculty Members who are Devoted to both Teaching and Research Excellence

Research prospers in favorable environments because supportive environments and good institutional governance retain talented researchers. For many traditionally non-research-intensive institutions, including some osteopathic medical schools, retaining faculty who are doing excellent jobs in both teaching and research has been becoming more and more challenging. Analysis of the failing schools in this regard indicates that a major reason is the lack of encouragement and respect for research excellence. To retain faculty of both teaching and research excellence, several measures outlined below have been demonstrated to be effective, at least to some extent, in many institutions of either research-intensive or teaching-oriented.

- Good governance making faculty feel that they are part of the institutional team of developing the biomedical research enterprise (see Chapter 9 for discussion of good governance). Avoiding micromanagement (see **Box 11-1** for definition of micromanagement) and fostering an environment of respect and trust so as to help maintain the high working enthusiasm of both faculty and supportive staff.

- Offering monetary incentives (salary supplement/bonus for those who are funded extramurally and who are doing an excellent job in both teaching and research).
- Providing appropriate teaching reduction to research-active faculty. However, it is not advisable to make them pure researchers. This is because of the following: (i) research active faculty are frequently also excellent educators; and (ii) the reality for the newly established osteopathic medical schools is that the institution prospers when its faculty do both excellent teaching and productive research.
- Providing reasonable amounts of resources (such as seed funds) to research-active faculty to allow them to develop new lines of research. Research requires long-term institutional investment, and the return is multiplicative (see Chapter 2).
- Providing graduate students supported by the institution, and fostering multidisciplinary graduate programs (see Section 5 below for discussion of multidisciplinary programs).
- Providing administrative support for grant application and management (see Chapter 6 for discussion of grant office).
- Emphasizing research priorities without compromising research freedom and intellectual creativity (see Section 4 below for more discussion of institutional research priorities).

Box 11-1. Micromanagement and its detrimental effects

Micromanagement means management with excessive control or attention on small details. Often, this excessive obsession with the most of minute details causes a direct management failure in the ability to focus on the major details. Case studies suggest that micromanagement by senior administrators at some medical schools may be detrimental to the development and prosperity of the institutions as reflected by the dampening of the working enthusiasms of the faculty experts and diminished motivation of the department chairs or program directors to produce innovative ideas and formulate constructive measures.

Micromanagement is considered a costly management style that often leads to the failure of an organization [9]. It is a manifestation of distrust and disrespect at institutions of higher education. Notably, micromanagement is also considered a fundamental mechanism of tumorigenesis [10].

4. Goal Three: Defining Biomedical Research Areas and Priorities to Develop Institutional Signature Research

Defining research areas and priorities is among the most crucial components of strategic biomedical research planning. This will allow the efficient use of institutional resources to maximize the research output and help build institutional signature research portfolio. Top medical research institutions always have well-defined research areas and priorities of their research portfolios, which are also dynamic as the biomedical field keeps changing from time to time. This section is intended to discuss the strategies for defining biomedical research areas and priorities at traditionally non-research-intensive institutions, especially osteopathic medical schools.

4.1. Strategy 1: To Encourage Research in "Osteopathic Areas"

As described in Section 4.2.2 of Chapter 10, the "osteopathic areas" of research overlap significantly with those of integrative medicine- the future of medicine and patient care [see References 11-15 of Chapter 10]. Research in these "osteopathic areas" is essential not only for strengthening the identity and uniqueness of osteopathic medicine, but also for further advancing the knowledge and practice of integrative medicine. In this context, the future of osteopathic medical research should not be underestimated. Resources (both internal and extramural funding) should be devoted to address the following two broad questions: (i) do they work? and (ii) how do they work? The research should focus on determining the clinical effectiveness and efficacy of the osteopathic modalities as well as on understanding the underlying biological mechanisms. In a broad sense, the osteopathic modalities that warrant extensive investigation include osteopathic manipulation treatment (OMT), mind-body medicine, exercise/physical activity, and dietary factors/natural products (**Figure 11-1**).

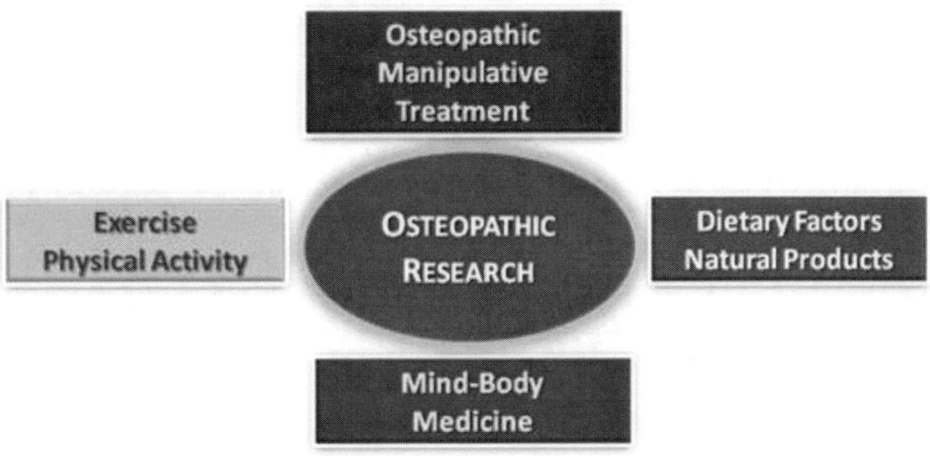

Figure 11-1. Osteopathic modalities that deserve extensive investigation to determine the clinical effectiveness and efficacy as well as the underlying biological mechanisms.

4.2. Strategy 2: To Encourage Research in "Non-osteopathic Areas" and Integrative Research

What makes a medical school a top one is not the type, but the quality and impact of the research it conducts. Although the "osteopathic areas" of research are very broad, researchers should not be restricted from doing the "non-osteopathic type" of research. Instead, researchers in osteopathic medical schools should be encouraged to conduct both "osteopathic" and "non-osteopathic" research with a particular emphasis on the combined effects of both osteopathic and non-osteopathic modalities. Research efforts in this regard will provide supporting evidence and rationale for practicing integrative medicine- the future of medicine and practice of patient care.

5. Goal Four: Stimulating Biomedical Research and Scholarly Productivity of Faculty to Maximize Institutional Research Output and Investment Return

5.1. Defining Research and Scholarly Activities

The research and scholarly activities (see **Box 11-2** for definition of scholarly activity) at institutions of higher education, especially osteopathic medical schools must be judged by multiple endpoints, which may include the following:

- Peer-reviewed publications in respected journals
- Other scholarly/professional activities (e.g., publishing books, serving as editors or editorial board members, serving on grant review panels or NIH study sections, and leadership in professional societies, etc) (see Table 6-1 of Chapter 6 for definition of NIH study section)
- Extramural grant funding levels, especially the amounts of NIH grant funding (see Chapter 6 for discussion of NIH grants)

The above research and scholarly/professional activities collectively contribute to the improvement of the visibility, prestige, and value of the medical college as well as the affiliated university as a whole. It is primarily for this reason that research and scholarly activities are considered as key credentials of the faculty and used as major criteria for faculty tenure and promotion and salary raise.

Box 11-2. Defining research and scholarly activities

Research and scholarly activities are terms frequently used in institutions of higher education and are increasingly important components of medical education though many are unsure about the exact difference between the two. The term research has been defined in various ways in this book (see Section 2 of Chapter 1). Research or research activity can be simply defined as a process or activity in which knowledge is created or advanced. On the other hand, the term scholarly activity refers to an oral or written presentation that reflects a thorough and critical collection of knowledge. Research activity and scholarly activity are intimately intertwined, and as such the compound term "research and scholarly activities" is often used to describe the scientific and professional contribution of faculty at institutions of higher education. Research and scholarly activities are also increasingly important components of medical education, including the training of family medicine residents [11].

5.2. Judging Research and Scholarly Activities at Traditionally Non-Research-Intensive Institutions

As osteopathic medical schools are usually non-research-intensive, researchers at these schools are usually less competitive than those at major research universities for securing extramural grant funding. As a result, federal funding (primarily NIH grants) to osteopathic medical schools is insignificant compared with that awarded to allopathic medical schools though NIH funding to osteopathic medical schools has increased over the past two decades (see Section 3.2.2.2 of chapter 2 and references 32-34 of Chapter 2). This reality necessities the allocation of a reasonable amount of internal funds to establish, support, and sustain biomedical research programs so that they progress and gradually become competitive for obtaining extramural grant funding. With that being said, judging the output and benefits of research investment in newly established osteopathic medical schools or traditionally non-research-intensive universities should not be primarily based on the levels of extramural grant funding. The value and impact of faculty peer-reviewed publications in respected scientific journals should not be underestimated. In fact, high quality publications have been used as one of the major criteria for the ranking of world-class universities (see Table 2-2 of Chapter 2). High quality research publications are also key to the success of grant applications to NIH and other federal and nonfederal funding agencies.

5.3. Strategies to Stimulate Research and Scholarly Activities

Due to the difference in research structure, faculty, and investment between research universities and traditionally non-research-intensive institutions, strategies to stimulate research and scholarly activities must be designed in a way that matches the reality of the institution. For teaching institutions, especially osteopathic medical schools, several strategies may be useful in facilitating faculty's research and scholarly activities without compromising their teaching responsibilities and excellence [7].

- Emphasizing good governance at the institutional level so as to maintain and augment the enthusiasm of faculty in achieving high goals of research and scholarly activities; avoiding micromanagement and encouraging participation of research-active faculty in decision-making and reaching consensus.
- Placing research and scholarly activities as one of the top priorities of the institutional mission statement, and forming a consensus among all members of the institution that faculty research and scholarly activities are essential to the improvement of the institution's prestige and value.
- To stimulate research and scholarly activities, faculty promotion and tenure will be based on publications and other scholarly activities and grant funding levels in addition to excellence in teaching and services.
- Providing monetary incentives and administrative/financial support so as to retain excellent faculty and augment their research productivity and scholarly activities. Case studies suggested that a high turnover rate of research-active faculty of teaching excellence is detrimental not only to the research enterprise, but also to the overall reputation of the institution.

6. Goal Five: Fostering Interdisciplinary Programs to Unify University Wide Research Enterprise

6.1. Defining Interdisciplinary Research

The definition of a discipline and discussions of the varieties of interdisciplinary, multidisciplinary, and trans-disciplinary research have occupied much scholarly debate. Although there is not always agreement on these definitions, it is clear that areas of research are dynamic- continually emerging, melding, and transforming. What is considered interdisciplinary today might be considered disciplinary tomorrow [12].

A working definition of interdisciplinary research is set forth in a National Academies' report [13], stating that interdisciplinary research is a mode of research by teams or individuals that integrates information, data, techniques, tools, perspectives, concepts, and/or theories from two or more disciplines or bodies of specialized knowledge to advance fundamental understanding or to solve problems whose solutions are beyond the scope of a single discipline or area of research practice. Important research ideas often transcend the scope of a single discipline or program. The value of interdisciplinary research in pushing the fields forward and accelerating scientific discovery has been widely recognized by both scientific communities and federal funding agencies, including the United States National Institutes of Health (NIH) and National Science Foundation (NSF) [12, 14]. It is also widely appreciated that the integration of research and education through interdisciplinary training prepares a workforce that undertakes scientific challenges in innovative ways [12]. This is particularly true for the biomedical science because development of biomedicine increasingly relies on technological advances in other scientific fields [15, 16].

6.2. Strategies for Facilitating Interdisciplinary Programs

Creation of interdisciplinary research and training programs has been becoming common practice at major research universities and top medical institutions. Because of the value of interdisciplinary programs in advancing both the biomedical research enterprise and improving the institutional prestige and value, developing such programs is also becoming a top priority at many traditionally non-research-intensive institutions, including some osteopathic medical schools. Interdisciplinary programs not only strengthen biomedical research at the osteopathic medical colleges, but also facilitate research progress at the affiliated universities. Together, they help unify university wide research (and education) enterprise to maximize institutional research (and education) output and investment return. Listed below are several strategies that may be used to facilitate the creation and development of interdisciplinary programs.

- Placing interdisciplinary programs as a top priority of institutional strategic research.
- Encouraging research-active faculty to establish collaborative research with productive researchers at other colleges of the affiliated university.
- It is essential that dedicated institutional fund be available for creating and developing interdisciplinary research programs.

- Establishing multidisciplinary graduate programs that attract bright graduate students who are motivated to do high quality research in areas involving multiple disciplines.

Conclusion and Perspectives

Establishing a successful research program in a traditionally non-research-intensive institution is one of the most challenging, yet important steps toward improving the prestige and value of the institution. This is particularly true for newly established osteopathic medical schools that usually lack sufficient resources and research-active faculty. Two areas required for success include securing financial support and attracting and retaining research-active faculty who are also devoted to teaching excellence. Since major research universities are struggling in securing extramural funding to maintain their research activities, it is important for newly established traditionally non-research-intensive institutions to set aside fund as a sustainable source of support (investment) for developing biomedical research. In this context, institutional intramural funding of research generally adds up to 20% of total research costs at major medical institutions [17], and this rate may be much higher at traditionally non-research-intensive institutions because of their relative lack of competitiveness for extramural funding.

Many academic institutions plan their investments in biomedical research largely on the basis of their anticipated success in competing for federal funds, predominantly those from NIH, a strategy that despite its risks, had worked relatively well for the past five decades until early/mid 2000's because of dependable public investment in the NIH [18]. However, this situation has changed drastically over the past several years due to economic constrains and budget cuts [18]. With the current apparently continued depression of NIH budget and the increasing competitiveness for obtaining extramural funding, institutional intramural funds are becoming more and more essential for maintaining and developing biomedical research especially at traditionally non-research-intensive institutions, including osteopathic medical schools.

In developing strategic biomedical research at traditionally non-research-intensive institutions, it is also important to remain realistic when setting up the goals and laying out the strategies and time frame to achieve the goals, as "Rome wasn't built in a day." In this regard, while the long-term research goal for a newly established osteopathic medical school is to build a nationally recognized research program, short- and intermediate-term goals must be realistic and should focus on establishing a unique, competitive, and productive research program that helps augment the institution's visibility and improve its academic status in a progressive, step-by-step manner. Of course, such a research program also benefits the education of medical students and their career as health care providers (see Table 10-1 of Chapter 10 for benefits of research to students' education). In this context, top medical schools provide excellent research opportunities to their medical students and encourage them to participate in both basic and clinical research. Without extensive exposure to medical research concepts or application, medical students may be generally under-prepared to efficiently consume and effectively apply research and evidence-based medicine information in patient care [19]. It is hoped that what have been discussed in this chapter as well as the preceding chapters will provide useful information for those who are involved in developing biomedical research at traditionally non-research-intensive universities to benefit the society as well as to increase the institution's prestige and value.

References

[1] Johns Hopkins Medical Institutions. Core facilities: Hopkins core conduit. http://www.hopkinsmedicine.org/Research/core_facilities/index.html (accessed on January 3, 2013).

[2] Harvard Medical School. Welcome to the Harvard catalyst core facilities database. http://cbmi.catalyst.harvard.edu/cores/index.html (accessed on January 3, 2013).

[3] NIH Office of Research Infrastructure Programs. Research facilities improvement. http://dpcpsi.nih.gov/orip/research_fimprov.aspx (accessed on January 3, 2013).

[4] Altbach PG, Salmi J. The Road to Academic Excellence: The Making of World-Class Research Universities (Directions in Development) (2011), the World Bank, Washington, D.C.

[5] Salmi J. Attracting talent in a global academic world: how emerging research universities can benefit from brain circulation. The Acaedmic Executive Brief. 2012; 2(1):1-5. http://academicexecutives.elsevier.com/sites/default/files/AEB_2.1_Salmi.pdf (accessed on January 3, 2013).

[6] Vicens Q, Bourne PE. Ten simple rules to combine teaching and research. *PLoS Comput Biol.* 2009; 5:e1000358.

[7] Kaplan K. Academia: small-school science. *Nature.* 2011; 477:239-41.

[8] Lindquist ES, Anderson LJ, Simmons JA. Teaching or research: small colleges aided by research networks. *Nature.* 2011; 478:458.

[9] Collins SK, Collins KS. Micromanagement: a costly management style. *Radiol Manage.* 2002; 24:32-5.

[10] Burgess DJ. Tumorigenesis: dangerous micromanagement. *Nat Rev Cancer.* 2010; 10:664.

[11] Recommended curriculum guidelines for family practice residents: research and scholarly activity. American Academy of Family Physicians (AAFP) Reprint No. 280 (revised June 2003). http://www.aafp.org/online/etc/medialib/aafp_org/documents /about/rap/curriculum/research_and_scholarly.Par.0001.File.tmp/Research.pdf (accessed on January 3, 2013).

[12] National Research Foundation (NSF). What is interdisciplinary research? http://www.nsf.gov/od/oia/additional_resources/interdisciplinary_research/definition. jsp (accessed on January 3, 2013).

[13] Committee on Facilitating Interdisciplinary Research, Committee on Science, Engineering, and Public Policy, National Academies. *Facilitating Interdisciplinary Research.* pp. 2, 2004, National Academies Press, Washington, DC.

[14] NIH. Interdisciplinary research. http://commonfund.nih.gov/interdisciplinary (accessed on January 3, 2013).

[15] Sharp PA, Langer R. promoting convergence in biomedical science. *Science.* 2011; 333:527

[16] Perzynski AT. Multidisciplinary approaches to biomedical research. *JAMA*. 2010; 304:2243-4.
[17] Boat TF. Insights from trends in biomedical research funding. *JAMA*. 2010; 303;170-1.
[18] Heinig SJ, Krakower JY, Dickler HB, Korn D. Sustaining the engine of U.S. biomedical discovery. *N Engl J Med*. 2007; 357:1042-7.
[19] Cruser dA, Dubin B, Brown SK, Bakken LL, Licciardone JC, Podawiltz AL, Bulik RJ. Biomedical research competencies for osteopathic medical students. *Osteopath Med Prim Care*. 2009; 3:10.

Self-Assessment Questions

[1] In the context of administrative activities at institutions of higher education, the term "Micromanagement" refers to:

A. Good governance
B. Management with excessive control or attention on small details
C. Participation of faculty, staff, and students in decision making
D. Tumorigenesis

[2] Which of the following is considered as an osteopathic modality that warrants further investigation to understand the underlying biological mechanism?

A. Acupuncture
B. Integrative medicine
C. Osteopathic manipulation treatment
D. Traditional Chinese Medicine

[3] Interdisciplinary research is a mode of research by teams or individuals that integrates information, data, techniques, tools, perspectives, concepts, and/or theories from two or more disciplines or bodies of specialized knowledge to advance fundamental understanding or to solve problems whose solutions are beyond the scope of a single discipline or area of research practice. Interdisciplinary research:

A. Is not supported by the U.S. National Institutes of Health
B. Is only supported by the U.S. Center for Disease Control and Prevention
C. Is only supported by the U.S. Food and Drug Administration
D. Is supported by the U.S. National Science Foundation

[4] A graduate student from a pharmacy school is working on a thesis project involving the use of a special material developed by a scientist at an engineering school of another university to deliver an old anticancer drug. To ensure the successful completion of the research project, the thesis committee consists of 2 scientists working in the material engineering field, one cancer biologist, and two professors from the department of pharmaceutical sciences of the pharmacy school. This research is best described as:

A. Engineering research
B. Multidisciplinary research
C. Pharmacy research
D. Translational research

[5] While building institutional research laboratories and other research infrastructure is usually supported by the institutional budget, federal funds are also available for such a purpose. For example, the U.S. National Center for Research Resources supports the Research Facilities Improvement Program (RFIP). The U.S. National Center for Research Resources is part of:

A. The U.S. Center for Disease Control and Prevention
B. The U.S. Food and Drug Administration
C. The U.S. National Academy of Sciences
D. The U.S. National Institutes of Health
E. The U.S. National Science Foundation

Answers to Self-Assessment Questions

Chapter 1: [1] A; [2] A; [3] B; [4] B; [5] D; [6] A; [7] D; [8] A; [9] D; [10] A; [11] A; [12] A; [13] C; [14] B; [15] D; [16] C; [17] B; [18] D; [19] C; [20] D.

Chapter 2: [1] D; [2] C; [3] D; [4] D; [5] A; [6] C; [7] A; [8] D.

Chapter 3: [1] D; [2] A; [3] E; [4] B; [5] D.

Chapter 4: [1] C; [2] B; [3] B; [4] D; [5] A; [6] B; [7] C; [8] C.

Chapter 5: [1] B; [2] C; [3] A; [4] C; [5] A.

Chapter 6: [1] C; [2] D; [3] B; [4] D; [5] D.

Chapter 7: [1] C; [2] B; [3] B; [4] B; [5] E.

Chapter 8: [1] B; [2] B; [3] D; [4] A; [5] B.

Chapter 9: [1] C; [2] B; [3] C; [4] D; [5] B.

Chapter 10: [1] A; [2] D; [3] C; [4] B; [5] D.

Chapter 11: [1] B; [2] C; [3] D; [4] B; [5] D.

Index

D

E

F

G

H

I

J

L

M

Q

R

S

T

U

V

W

Y